50 PIC Microcontroller Projects
for beginners and experts

50 PIC Microcontroller Projects
for beginners and experts

Bert van Dam

Elektor International Media BV
Postbus 11
6114 ZG Susteren
The Netherlands

All rights reserved. No part of this book may be reproduced in any material form, including photocopying, or storing in any medium by electronic means and whether or not transiently or incidentally to some other use of this publication, without the written permission of the copyright holder except in accordance with the provisions of the Copyright, Designs and Patents Act 1988 or under the terms of a licence issued by the Copyright Licensing Agency Ltd, 90 Tottenham Court Road, London, England W1P 9HE. Applications for the copyright holder's written permission to reproduce any part of this publication should be addressed to the publishers.

The publishers have used their best efforts in ensuring the correctness of the information contained in this book. They do not assume, and hereby disclaim, any liability to any party for any loss or damage caused by errors or omissions in this book, whether such errors or omissions result from negligence, accident or any other cause.

British Library Cataloguing in Publication Data
A catalogue record for this book is available from the British Library

ISBN 978-0-905705-88-0

Prepress production: Kontinu, Sittard
Design cover: Helfrich Ontwerpbureau, Deventer
First published in the United Kingdom 2009
Printed in the Netherlands by Wilco, Amersfoort
© Elektor International Media BV 2010

109010-UK

INTRODUCTION	8
1 WHAT IS A PIC MICROCONTROLLER?	9
2 WHAT YOU WILL NEED	11
2.1 NECESSARY ITEMS	11
2.2 OPTIONAL ITEMS (NICE TO HAVE)	22
3 TUTORIAL PROJECT	29
3.1 THE HARDWARE	29
3.2 THE SOFTWARE	35
3.3 COMPILING AND DOWNLOADING	39
3.4 DEBUGGING	41
3.5 DONE!	43
3.6 OTHER RESULTS	43
4 RELAY	45
4.1 AUTO DISENGAGING PIC	45
4.2 BISTABLE RELAY	50
4.3 FLASHING LIGHT (8 VOLTS)	56
4.4 FLASHING LIGHT (MAINS, 110 OR 240 VOLTS)	59
5 ALTERNATING CURRENT	62
5.1 SWITCH (8 VOLTS)	62
5.2 ZERO CROSSING DETECTION	66
5.3 LIGHT DIMMER (8 VOLTS)	71
5.4 LIGHT DIMMER (MAINS, 110 TO 240 VOLTS)	78
6 MAKE SOUND	82
6.1 YOUTH DETERRENT	82
6.2 DIGITAL TO ANALOG (D/A)	92
6.3 SINUS FROM A LOOKUP TABLE	100
6.4 SIREN WITH AMPLIFIER	109
6.5 A TALKING 18F4685	114

7 PROCESS SOUND 123

    7.1 Comparator   123
    7.2 Sound switch   129
    7.3 Artificial Ears   135
    7.4 Frequency meter   139
    7.5 Microphone pre-amplifier   148

8 SENSORS 150

    8.1 Hall effect object protection   150
    8.2 Touch key   154
    8.3 Capacitive (no contact) level gauge   156
    8.4 Low voltage alarm   162
    8.5 Temperature control   168
    8.6 Temperature in a poultry farm   171

9 COMMUNICATION 183

    9.1 RS232 - Passthrough communication   183
    9.2 RS232 - VT52 terminal   189
    9.3 IR - Receiver   194
    9.4 IR - Transmitter (remote control)   203
    9.5 USB - Serial echo   213
    9.6 USB - Teasing mouse   223
    9.7 USB - A/D measurements in Excel   229
    9.8 CAN bus - Loopback   237
    9.9 CAN bus - Remote LED   246
    9.10 SPI - Master - slave   253
    9.11 SPI - Sampling to an MMC card   263
    9.12 I^2C - Real Time Clock (RTC)   270
    9.13 I^2C - Egg timer   281
    9.14 I^2C - Memory with a back-up battery   284
    9.15 I^2C - Eight pin I/O expander   286
    9.16 I^2C - D/A conversion   297

10 CAMERA VISION 303

    10.1 WHERE IS MY PAPER? 311
    10.2 COUNT THE COLORED SQUARES 324
    10.3 I BELIEVE SOMETHING HAS CHANGED... 330
    10.4 MAKING PICTURES FOR YOUR PC 336

11 MISCELLANEOUS 342

    11.1 SEVEN SEGMENT DISPLAY 342
    11.2 TWO 7-SEGMENT DISPLAY'S WITH TRANSISTOR SWITCHING 346
    11.3 ROTARY ENCODER 351
    11.4 PORT B INTERRUPT 356
    11.5 UPGRADE YOUR WISP PROGRAMMER FIRMWARE 360
    11.6 LASER ALARM 364

12 OTHER MICROCONTROLLERS 368

    12.1 SUPPORTED MICROCONTROLLERS 368
    12.2 MIGRATION 371
    12.2.1 HOW DOES IT WORK 371
    12.2.2 Case 1 - from a 16f877A to a 10f200 (purpose: reduce cost) 371
    12.2.3 Case 2 - from a 16f877A to a 18f4455 (purpose: add USB) 375

13 APPENDIX 377

    13.1 JAL 377
    13.1.1 General 377
    13.1.2 Syntax 378
    13.2 LIBRARY _BERT 396
    13.3 OTHER LIBRARIES 404
    13.4 ASCII TABLE 415
    13.5 KEYBOARD SCANCODES 417
    13.6 TRANSISTOR 419
    13.7 CONTENT OF THE DOWNLOAD 423
    13.8 TIPS AND TRICKS 427

Index 431

Introduction

This book covers a series of fun and exciting projects. From a simple flashing LED to a camera vision project, from laser alarm to USB mouse, from capacitive level gauge to mains light dimmer.

You can use this book as a projects book, and built and use the 50 projects that it covers. The clear explanations, schematics and pictures of the project on a breadboard make this a fun activity.

You can also use this book as a study book. In each project the technical background is covered in detail, explaining how and why this project works. We will also discuss how to use the datasheets. As you go along you learn more and more about the projects and the microcontrollers, and you will be capable of expanding each project, or adapt it to your own needs. For that reason the range of microcontrollers used in this book is limited to the 16f877A, 18f4455 and 18f4685.

Chapter 12 covers a few other microcontrollers. Fifteen in total, from small (10f200) to large (18f4685). Particularly the migration of your program from one microcontroller to another is discussed.

Apart from a projects or study book this book can also be used as a reference guide. The explanation of the JAL programming language and all of the expansion libraries used in this book is unique and nowhere else to be found. Using the index, you can easily locate projects that serve as examples for the main commands. Even after you've built all of the projects in this book it will still be a valuable reference guide to keep next to your PC.

At this point I would like to thank Wouter van Ooijen for his support many years ago when I started with microcontrollers myself, and for his help with the editing of this book. And also Bert Oudshoorn for his help with testing the projects and with the final editing. Of course Kyle York, for the incredible amount of work he has done in keeping the JAL compiler updated, and his help with the English translation. And finally the hundreds of people with whom I've been in contact in the JAL group or by e-mail and who contributed to the libraries or software.

I hope you, as my readers, will enjoy the wonderful world of microcontrollers as much as I do.

Bert van Dam

1 What is a PIC microcontroller?

PIC^{TM} is a trademark of the Microchip Corporation. It's not quite clear whether this is just a name or an abbreviation. On the Internet names like Peripheral Interface Controller or Programmable Integrated Controller are occasionally used, but on the website of the manufacturer PIC^{TM} is used as if it is a normal name which just happens to be spelled in capitals.

A PIC is actually a complete mini computer on a chip. As opposed to a normal computer, such as a PC - which is an abbreviation and stands for Personal Computer - a PIC is not designed to work with people but with machines.[1] This means there is no simple way to connect it to a keyboard or a terminal. There are however many ways to connect it to machines or parts of them, such as switches, LEDs, variable resistors, temperature sensors, infrared sensors, or even other microcontrollers.

Figure 1 PIC microcontroller (16f877A)

Because they can be programmed microcontrollers can be used in many different ways. You can find microcontrollers in VCRs, remote control units, cars, vending machines... they control motors and heaters, decipher remote control signals, measure temperatures, and much more. It is this reason of being found inside machines that microcontrollers are often called "embedded systems".

---

[1] There is another important but very technical difference. In a PC both program and data occupy the same memory. This is called the von Neumann architecture. The advantage is that there is a flexible relationship between the available memory for the program and the data. In a PIC™ separate memory is used for the program and the data, the so called Harvard architecture. The advantage is that program and memory can be accessed at the same time, which significantly improves the processing speed.

You can design and build applications like these embedded systems yourself. With a few simple parts and the instructions in this book you can build some very interesting projects. Each project comes with a clear and complete explanation of both the software and hardware. Perhaps even more important is that for each project the technical background is explained in detail. This allows you to understand how and why these projects work, so that you can easily adapt them to your own needs and ideas.

The book starts very simply with a tutorial project and step-by-step instructions. As you go along the projects increase in difficulty and only the new concepts are explained. By the end of the book you will no longer be a beginner!

Chapter 13 contains a complete and unique overview of all of the commands of the JAL programming language and the expansion libraries used. You won't find this anywhere else! You can use it as reference guide, but also to find fun and useful commands and options.

All of the software used in the book is explained and shown. To avoid having to type in the programs yourself, they are included along with the other software needed in a complete package, which you can download for free at http://www.boekinfo.tk.

To get started quickly, a hardware kit is also available. It contains all of the parts needed for a selection of projects. The kit is designed for your convenience, but of course you can also obtain and use your own parts. More information on ordering and international shipping can be found at http://www.boekinfo.tk.

All of the projects in this book are built on a breadboard. This allows you to easily build and expand them before you design a printed circuit board. It is more fun to see the projects in this book as a starting point for something that suits your own needs rather than finished projects.

Okay, enough talk... let's get started with our first PIC microcontroller project!

2 What you will need

This chapter covers all of the materials needed to start with this fascinating hobby. And there are many different options to choose from. Other choices may work just as well as the ones chosen here. What's important is that the selected materials work together perfectly. The Wisp648 programmer, Xwisp, JALedit IDE, JAL and standard _bert libraries are just such a combination; a perfect fit, giving you many hours of fun.

Of course you also need this book - preferably your own copy - so you can make notes and even write in the book. Your own experiences and notes about the parts that you use will make this book even more valuable. Writing in a new book for the first time may not be easy, but it gets easier in time.

A lot of the software (at least the most important software) can be downloaded from the website http://www.boekinfo.tk in the section "download packet". You'll also find information on a hardware kit for beginners with this great hobby, links to suppliers, and lots of other information.

2.1 Necessary items

Breadboard

Unless you're the type that goes for the soldering iron immediately it is a wise idea to use a breadboard. This will allow you to build and modify projects quickly, making it ideal for prototyping.

Figure 2 Breadboard with decoupling capacitors

To prevent power issues, decoupling capacitors of 0.1 uF are inserted on the power strips at the corners. In one of the corners a stabilizing capacitor of 22 uF is used, as well as a small LED (reverse mounted, so not lit), which serves as a limited safety diode against an accidental reversed power connection. You must use this set-up in all projects in this book. For that reason we will not mention them at each separate project.

Depending on what you want to build you will need electronic parts. If possible try to get parts that fit into a breadboard. Some parts may have wires that are too thick. Don't force them into the breadboard, but solder small wires to them instead, otherwise you will bend the small contacts inside the breadboard.

Many suppliers sell small bags of mixed parts (resistors, LEDs, capacitors). This may be a good way to start your parts collection.

Long breadboards such as the one in Figure 2 have a power rail that is split into two parts. These are not necessarily internally connected. The top and bottom rails are almost never internally connected. So, before you start a project it's best to prepare the breadboard by connecting all the power rails and inserting the parts described previously. You wouldn't be the first to wonder why a project doesn't work only to find out later that part of the power rail you used was not connected to the power source.

Power source (UA7805)

All PIC microcontrollers used in this book use 5 volts, and that has to be rather accurate, especially during programming. If you use the recommended Wisp648 programmer you will not need a separate power supply so you can skip this section. Otherwise use a good stabilized power source with a UA7805 (or similar[2]) active stabilizer, connected according to the schematic below.

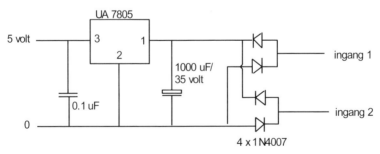

Figure 3 Stabilized 5 volts power supply.

---

[2] For example the LM7805 manufactured by Fairchild Semiconductor. This power regulator can handle 1 Amp max, while the UA7805 manufactured by Texas Instruments can handle 1.5 Amp.

A transformer is connected to feed 1 and feed 2. This may very well be a wallwart power supply, as long as the voltage is high enough. Preferably 9 to 24 volts if the transformer is alternating current (AC), or 9 to 32 volts if the transformer is direct current (DC).

It doesn't make any difference how you connect a DC transformer, since the voltage is rectified by the four diodes. Of course the transformer needs to have sufficient capacity. How much "sufficient" is depends on your project. Several hundred milliamps is usually enough. If you use more power it is advisable to equip the UA7805 with a heat sink to provide cooling[3].

Programmer (Wisp648)

Once you have written a program it needs to be transferred into the microcontroller. This is the task of a programmer. Some programmers can program a microcontroller while it is still in the circuit on the breadboard. This is called "in-circuit programming". Trust me, you really, really, REALLY want this feature. It gets annoying very quickly when you have to wriggle the microcontroller off the breadboard each time you forgot a comma in your program.

We'll be using the Wisp648, an intelligent programmer than can handle a wide array of different PIC microcontrollers. The intelligence of this programmer is, by the way, contained in its own microcontroller; so you can make your own updated version of the program software if needed.

Also important is that this programmer has a pass-through feature. Without touching as much as a single wire you can communicate from your PC directly with the microcontroller. This makes it ideal for debugging your software. You will be using this functionality quite often in this book.

In order to use the microcontrollers described in this book your Wisp needs firmware version 1.28 or newer. If your Wisp has an older firmware version you need to perform a free upgrade as described in section 11.5.

The Wisp648 is powered by a wallwart power supply, preferably 9 to 18 volts DC.

---

[3] For the projects in this book a heat sink of about 6 cm^2 is sufficient

Figure 4. The Wisp648 in-circuit programmer

Microcontrollers (PIC)

Which microcontroller to choose is a difficult question. There are dozens of different types, all with different options and cost. In this book the majority of the projects will use three PIC microcontrollers, the 16f877A, 18f4455 and 18f4685.

These microcontrollers are filled to the rim with options, and they have a lot of pins. Having many pins is an advantage because it means that pins have less combined functionalities, which is always a difficult subject for beginners. Besides they are popular microcontrollers which means that there is a lot of information (and thus help) to be found on the internet.

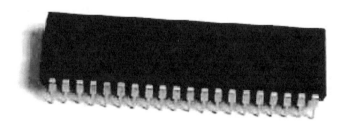

Figure 5. The 40 pin 16f877A microcontroller.

In chapter 12 we will show you how to migrate software from one microcontroller to the other. These are some of the specifications for the three microcontrollers used in this book:

Item	16f877A	18f4455	18f4685
Program memory	8192 words	12288 words	49152 words
RAM memory	368 bytes	2048 bytes	3328 bytes
EEPROM memory	256 byte	256 bytes	1024 bytes
I/O pins	33	35	36
Analog inputs	8	13	11
RS232	yes	yes	yes
USB	no	yes	no
I^2C	yes	yes	yes
PWM	yes	yes	yes
CANbus	no	no	yes
Speed	5 mips	12 mips	10 mips

Table 1. Specifications of three microcontrollers.

When you order this microcontroller you will need to know which package you want. This doesn't refer to the box the microcontroller is shipped to you in, but to the external connections of the chip. For instance, whether the microcontroller should have long pins that fit into a breadboard or printed circuit board, or small pins for surface mounting. The package shown in Figure 5 is called PDIP (or Plastic DIP), and this is what we'll be using.

Programming language (JAL)

Many different languages for programming microcontrollers exist on the market. The most commonly used language is arguably assembler (ASM) - a very powerful but extremely cumbersome language. Even the simplest program in assembler contains a long list of commands... definitely not suitable for beginners.

This book uses JAL (Just Another Language), a Pascal-like high-level language. It is the only advanced language that is completely free, and has a large international and very active user base. JAL is configurable and expandable using libraries, and can even be combined with assembler. This allows you to incorporate ASM snippets[4] found on the Internet.

This is an example of a JAL program that flashes an LED connected to pin 23 (c4):

```
-- JAL 2.4j
include 16f877a_bert

-- define variables
var bit flag

-- define direction of the signal of this pin
pin_c4_direction = output

forever loop

   -- change pin 23 status
   flag = ! flag
   pin_c4 = flag

   -- wait 100 milliseconds
   delay_100ms(1)

end loop
```

JAL has a very active usergroup with participants from all over the world. The address is listed on the website www.boekinfo.tk

Library (_bert)

Libraries contain additional functionality that can be added to JAL. You can, for example, make serial connections, read an analog signal, and much more.

In this book the _bert[5] standard libraries are used, combination sets of the most often used libraries. These libraries are combined into a perfect combination that allows for very easy programming, and facilitates the migration of software to other microcontrollers

---

[4] Small pieces of program that do something "convenient" and are exchanged between users.
[5] A silly name for a library. When I made the first one many years ago I called it _bert to distinguish it from the libraries I got off the Internet. The library became much more popular than I imagined, and since they are so well known I can't get rid of the name anymore!

with _bert libraries. Use of these _bert libraries also means that all software in this book is compatible with my previous JAL books ("PIC Microcontrollers - 50 projects for beginners and experts" and "Artificial Intelligence - 23 projects to bring your microcontroller to life!)

Functionality (with the standard library)	_bert library
Serial communication - hardware	x
Serial communication - software	x
Pulse Width Modulation (PWM)	x
A/D conversion	x
Access to Program memory	x
Access to EEPROM memory	x
ASCII	x
Delay	x
Registers and variables	x
Random number generator	x

Table 2. Functionality of the _bert library.

Because some of the functionality of the microcontroller is engaged by default by the _bert libraries some of the pins are reserved for these functions as shown in the next table. If you want to uses these pins for other functionalities you need to change the library. In practice you will virtually never need to do this.

This is the standard configuration:

16f877A pins	18f4455 pins	18f4685 pins	function
13 and 14	13 and 14	13 and 14	crystal
16	16	16	pwm (2)
17	17	17	pwm(1)
25	25	25	rs232 hardware TX
26	26	26	rs232 hardware RX
39	39	39	rs232 software RX
40	40	40	rs232 software TX
2 to 5 and 7 to 10	2 to 5 and 7 to 10	2 to 5 and 7 to 10	analog input (adaptable)
others	others	others	digital in/out

Table 3. Default configuration for three microcontrollers.

The analog pins 2 to 5 and 7 to 10 can be switched to digital using a command from the library. The free download available with this book adds a whole range of libraries to the JAL language. Most of these are discussed in projects in this book. Different people have written these libraries. You can see this by opening a library and reading the comments. Credit for these libraries goes of course to the writers. The functions in these libraries are also shown in the appendix. A unique overview that you will not find anywhere else!

Editor (JALedit)

You can use JAL from the command line, but an editor is much more convenient. An example of such an editor is JALedit. This freeware editor has a number of unique advantages:

- JAL commands are color coded.
- verview of variables, constants, and procedures used in a program.
- Block comment add or remove.
- One click compile and download.
- Built in ASCII table, calculator, and terminal.
- Source export to html (for publication on a website).
- Automatic backup.
- Compiler output in a separate window.

- Load multiple files at the same time.
- Jump to offending program line on error.

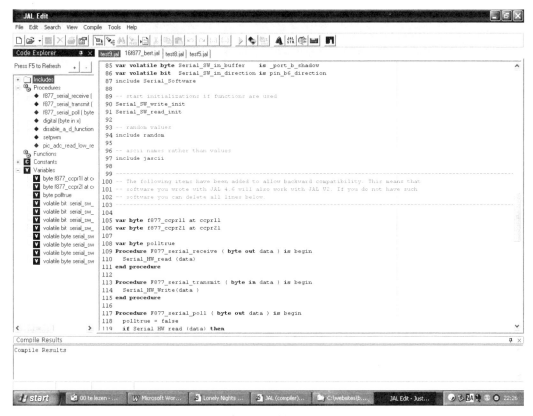

Figure 6 JALedit

This beautiful editor is part of the download package. If you plan to use this editor you can jump to the next item in this section.

If you prefer to use your own editor because you are used to it or because you find it more convenient that is of course also possible. If your program is called MyProgram.jal you can use the following command to compile this program to a HEX file:

C:\PICdev2\JAL\Compiler\jalV24j\ JALv2.exe MyProgram.jal -Wno-all -long-start -clear -s c:\picdev2\jal\libraries

The command should be on a single line when you use it, but that didn't fit in the book. The result of this command is a HEX file called MyProgram.hex. You need to use the Xwisp program to send this file to the programmer, with the following command

C:\PICdev2\xwisp\xwisp.exe port com1 wait err go MyProgram.hex

If your Wisp programmer is not connected to port 1 you need to change com1 to the correct port number. Note that there is no space between com and the number.

Xwisp has its own GUI (Graphical User Interface) to facilitate downloading hex files into the microcontroller, so you can use that instead of the previous command line. This is also convenient when someone sends you a hex file that you want to download into the microcontroller. You can start the GUI by double clicking on xwisp_gui.exe in the C:\PICdev2\xwisp directory. In the background a black window will open. This is normal, it will close automatically then you exit the GUI.

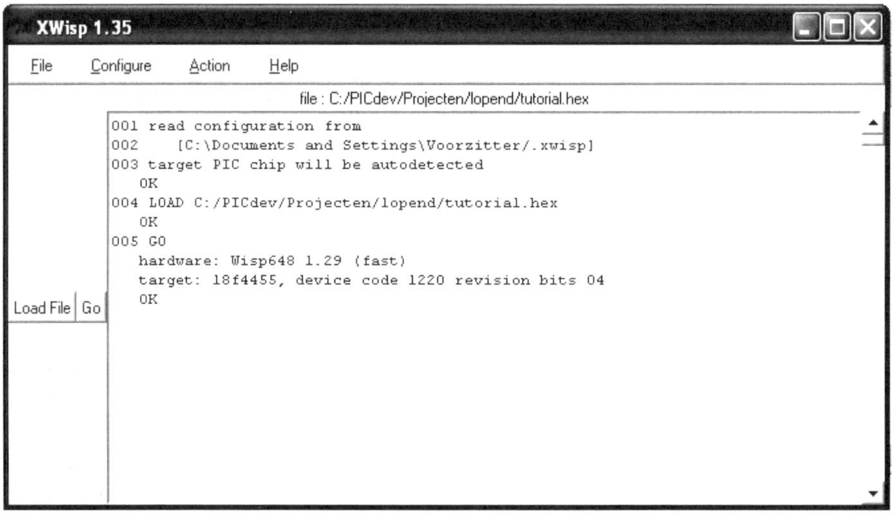

Figure 7. The Xwisp GUI.

Terminal software (MICterm)

In many projects in this book a serial connection is used to display data on a PC. That same serial connection is also a very convenient tool for debugging.

You could use a standard terminal program, but a much easier way is to use something written especially for microcontrollers: MICterm. This freeware program has a number of unique features:

- Switches the programmer automatically to pass-through.[6]
- Data can be shown raw, in hex or in ASCII.
- Data can be displayed in any combination of numbers, binary, on a dial or in a graph (with average). The speed of the graph can be changed, as well as the range.
- Separate window for send and receive
- Optionally send files to, or receive files from, the microcontroller.
- Send complete strings or single bytes.
- Settings are saved on exit.

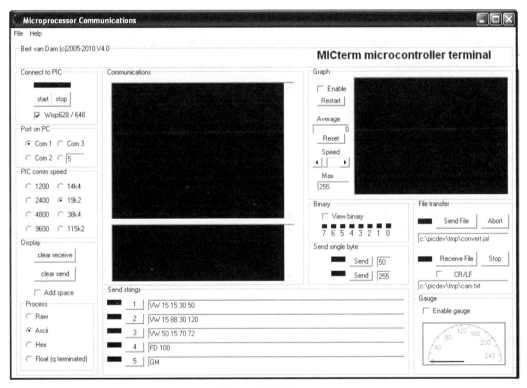

Figure 8 MICterm

MICterm is part of the download package.

---

[6] For Wisp programmers only.

2.2 Optional items (nice to have)

The following items are not necessary to work with microcontrollers, but they are very convenient.

Development software on the PC (Visual Basic)

Since the microcontroller can communicate with the PC it's fun to write special software for this. It is not necessary, since you can do all the projects in this book with the supplied software in the download package.

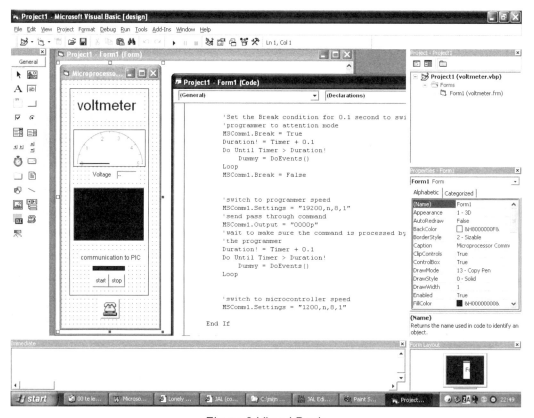

Figure 9 Visual Basic

If you own Visual Basic, for example, you could write beautiful applications for your projects such as a voltsmeter, several graphs on one page, and visualization of measurements. Many PC applications in the download package have been written in Visual Basic, and in many cases the source code is included. You can of course use another language if you want. Visual Basic[7] is commercial software, so it's not part of the free download package.

Oscilloscope (Software)

For power spikes and ripples, and projects with inexplicable behavior, sometimes it's very convenient to have a simple oscilloscope.

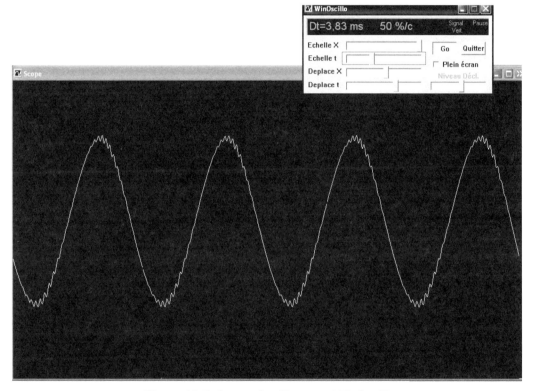

Figure 10 Software Oscilloscope WinOscillo

The software oscilloscope WinOscillo uses the soundcard of a PC to convert the signals. One of its limitations is that you need to make a small interface to protect the soundcard.

---

[7] Visual Basic is supplied by Microsoft. See http://www.microsoft.com

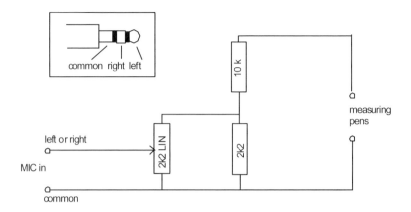

Figure 11. Interface for mic or line input of a soundcard on a PC.

The interface in Figure 11 is used to reduce the maximum voltage on the microphone, or line input of the soundcard, from 5 volts to 0.9 volts. Never connect the interface to voltages over 5 volts. Make sure you connect the common and left or right wires correctly, and make sure that the common wire is connected to the ground of your project, and not the other wire. Inside your PC the common of the soundcard, and the ground of RS232, USB are all connected. So if you switch those around on your project you are likely to blow a fuse or cause some damage.

In the Windows Mixer you need to enable the microphone, and turn the level to full. Switch off noise suppression and other signal enhancements if the Mixer shows these features.

Apart from just viewing the signal WinOscillo can make a recording of the signal, and take a snapshot of the image so you can save it. In this book you will find several examples.

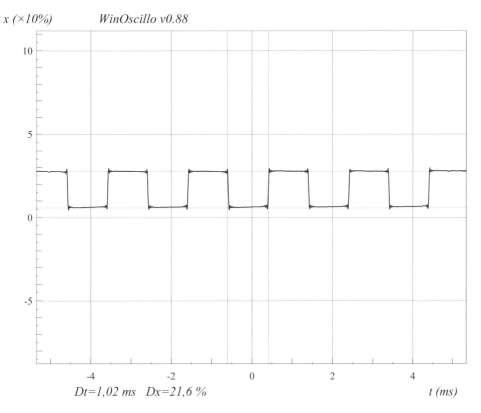

Figure 12. WinOscillo saved image with measuring lines and timed interval.

Unfortunately the user interface of WinOscillo is in French. The next list shows the commands that you will use most, with the English translation. Fortunately you will only need a few, and the shortcuts are easy to remember.

F2	Select oscilloscope...
F3	... or spectrum.
alt-enter	Full screen (with controls on top).
Ctrl-S	Store signal (as a short .wav or as .csv for use with Excel).
Ctrl-L	Load signal (select pause first otherwise it will take ages!).
Ctrl-M	Take a snapshot as .wmf.
Ctrl-V	Vertical cursor line.
Ctrl-H	Horizontal cursor line.
Crtl-W	Switch between both (if both are enabled).

Crtl-G	Tone generator.
echelle	Scale.
deplace	Move.
niveau decl.	Trigger level (if engaged).
Dt	Horizontal time.
%/c	Percentage of the maximum of a single on-screen square.

The two cursor lines can be moved using the mouse. The striped line can be moved with the left mouse button, the dotted line with the right mouse button. The lines will jump to the mouse cursor so you can set the mouse "just right" and then press the mouse button. Which lines are active is shown in the window (Hor. or Vert.)

Since WinOscillo uses the soundcard to collect the signal the frequency range is rather limited (approximately 20 Hz to 20 kHz). But it is still a very useful tool for microcontroller application developers. This freeware program is part of the download package.

Resistor and capacitor codes

Resistors and capacitors usually have a (color) code to indicate their value. This value can be found in tables, but you didn't buy a PC to fiddle around with paper tables. Two small programs can do the job for you:

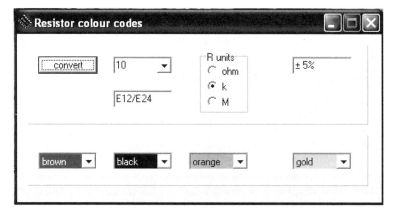

Figure 13 Resistor color-coding.

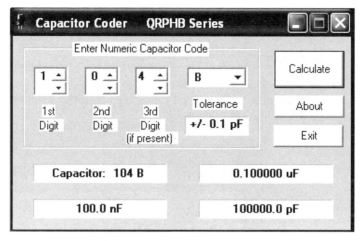

Figure 14 Capacitor coding.

Frequency analyzer (Software)

When you are working with interrupts or sound a frequency analyzer can be very convenient. The software frequency analyzer, or Spectrum Analyser, used in this book uses the soundcard of your PC, and thus the same interface as WinOscillo. With respect to frequency it also has the same limitations, approximately 20 Hz to 20 kHz. This freeware program is part of the download package.

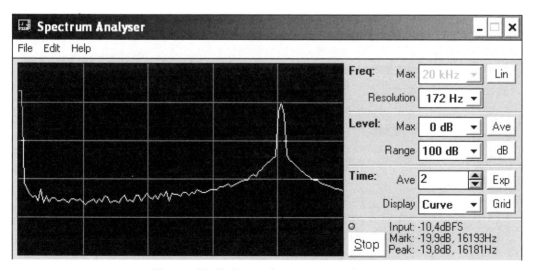

Figure 15 Software frequency analyzer.

Frequency generator

In several projects, such as 7.4, you need sound with a certain frequency. Of course you can use a microcontroller to built your own frequency generator, but sometimes it's easier to use a PC based software frequency generator.

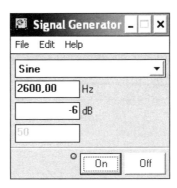

Figure 16. Software frequency generator.

This small generator can make a wide range of signals, with different frequencies. The sound is generated on the sound system of your PC. In order to use it in your project you need a plug that fits into the speaker outlet and ends in two wires that can be inserted in a breadboard. Again make sure to connect the ground wire to the ground of your project. This freeware program is part of the download package.

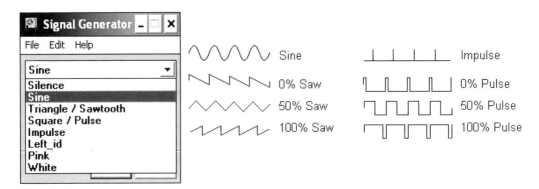

Figure 17. The generator can make many different signals.

3 Tutorial project

Now that all of the materials have been acquired it's finally time to get started! The purpose of this tutorial project is to build a set of flashing LEDs. In preparation it may be convenient to read Chapters one and two, but if you'd like to move right along that's no problem.

In this tutorial it is assumed that you have downloaded and properly installed the free software package from the website http://www.boekinfo.tk. The package contains the JAL programming language and the libraries, so you cannot continue without installing this first.

The package also contains an editor (JALedit) specifically designed for use with JAL. You can use your own editor (and if you don't have Windows on your PC you may even have to), but you'll need to follow the instructions that came with it. You might miss certain features such as syntax color-coding, single click compile and download, library view, and the procedure/function overview.

It is also assumed that you use the recommended programmer Wisp648 that you can purchase separately. You can use any other programmer (as long as it is capable of programming the microcontrollers used in this book), however you must also follow the instructions that came with it. And, again, you might not be able to use all of the functionality assumed in the book, such as in-circuit programming and RS232 pass-through.

3.1 The hardware

First of all, a microcontroller must be selected. Normally that would be the smallest and cheapest microcontroller that just meets all of the demands. For this project, however, we will select an easy microcontroller, which can be used to build several of the projects in this book: the 16F877 from Microchip. It is an easy to use microcontroller with many features.

The 16F877A has 40 pins. What these pins can do is listed in its datasheet.[8] Don't let the word "datasheet" fool you; it's more like a book of over 200 pages! In the datasheet you'll find the following drawing of the pin configuration.

---

[8] You can download the datasheet for free from Microchip at http://www.microchip.com. The datasheet name will resemble 16F87X, since similar chips are often combined into a single document.

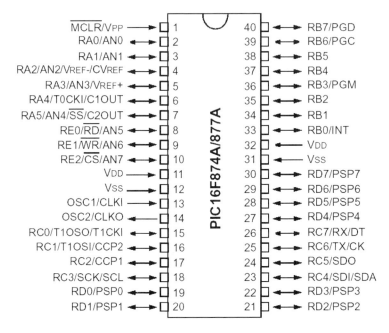

Figure 18 Pin layout 16f877A

Shockingly complex, but perhaps adding a few words will simplify matters a bit.

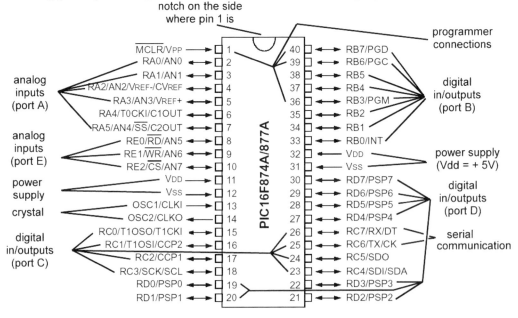

Figure 19 Pin layout 16f877A with explanation

What this all means:

- Analog input means that these pins can accept an analog signal, so a signal anywhere between 0 and 5 volts (for example, 3.56 volts).

- Digital in-/output means that these pins can process a digital signal...either 0 or 5V, but nothing in between. Since they are also outputs the pins can be made 0 or 5V by the microcontroller itself.[9]

- Power is obvious, where V_{SS} = 0V or ground, and V_{DD} = +5V. Beware that all power pins must be connected!

- Crystal is where the crystal (with a few capacitors) must be connected. The crystal takes care of a very stable (electrical) vibration, which is used to accurately control the speed of the microcontroller. This is called the (external) clock. All PIC microcontrollers can do without a crystal, but the speed will be lower and may be not accurate enough for communications with another microcontroller or a PC.

- Serial connection means that with these pins a serial (also called RS232) connection to another microcontroller or a PC can be made.

A lot of pins have multiple abbreviations listed next to them. This means that they have multiple uses. Some of these uses will be addressed in this book. For the time being it's enough to remember which pins are analog and which are digital.

Now that we know what the pins are for the hardware can be built. If you purchased the hardware kit you can start immediately. Otherwise, you'll have to shop for parts. That means you need to read the entire project, and use the parts mentioned in the schematic as a shopping list.

Since this is a tutorial project the parts for this project are listed here for your convenience:

---

[9] The actual value of a digital signal is a bit more complex. For example, 4.8 volts is still regarded as "on" and 0.5 volts is still regarded as "off".

Part	Type and number
PIC	16f877A
Resistors	2 x 330 ohm (orange-orange-brown)[10] 1 x 10 k ohm (brown-black-orange) 1 x 33 k ohm (orange-orange-orange)
LED	2 pieces
Capacitors	2 x 20 pF (code 20) 5 x 100 nF (code 104)
Crystal	1 x 20 MHz
Miscellaneous	1 breadboard 1 programmer, preferably Wisp648 wire (22 gauge or 0.5 mm^2 Cu) if you don't use the Wisp648: a stabilized 5V power supply

Table 4. Parts for the tutorial project.

Pins 1, 11, 12, 31, and 32 must always be connected, regardless of the project you are making. When a crystal is being used, pins 13 and 14 must also always be connected.

The only thing we need to choose for this project is where to connect the LEDs. And since we can choose any pins of ports B, C or D, we'll just pick two at random: 22 (d3) and 23 (c4).

The components on pin 1 are mandatory for programming the microcontroller. In many circuits you may find just a 33k resistor on this pin, which works fine too. The crystal is also required and is connected with two capacitors to pin 13 and 14.

---

[10] In resistor and capacitor values you often find letters. These have the following meaning: k=1,000; m=1/1,000; u=1/1,000,000; n=1/1,000,000,000, p=1/1,000,000,000,000. The location of the letter is also important. For example, 22k means a resistor of 22,000 ohms, but 2k2 means a resistor of 2,200 ohms. Sometimes the letter R is used for a decimal point: 4R7 would indicate 4.7 ohms.

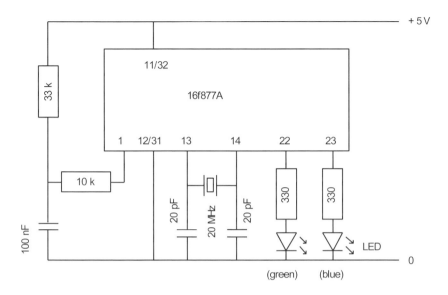

Figure 20 Schematic of the tutorial project

In this example the LEDs are blue and green. But the color is not important. The resistor in series with each LED, however, is very important. It limits the current through the LED. Without this resistor the microcontroller or the LED would break down - and it's usually the most expensive component that breaks first.

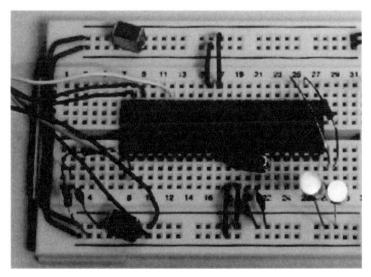

Figure 21 Hardware of the tutorial project

Place all of the components on the breadboard. Be certain that all wires are actually inserted into the holes, and that all 40 pins of the 16F877A are in the holes as well and not bent underneath the microcontroller. On one end of the microcontroller you'll see what looks like a half moon; this is where pins 1 and 40 are located. Often, there is also a small circle to indicate pin 1. Follow the schematic exactly, and use the picture in Figure 21 to verify that you understand the schematic correctly. On the left-hand side you can see the wires that connect the top and bottom power rail, and the long wires from the programmer. These are never in the schematic, but you need to connect them anyway.

On each of the four corners of the breadboard is a 0.1 uF (which is the same as 100 nF) capacitor. (Note that in the picture only two corners can be seen.) These take care of power ripples.

The long wires leaving the picture on the left are from the programmer. If you use the Wisp648 in-circuit programmer the connections must be made as follows:

16f877A	
color	pin number
yellow	pin 1
blue	pin 40
green	pin 39
white	pin 36
red	+ 5 volts
black	0
jumper	no

Table 5. Connection of the Wisp programmer.

The red and black wires of the Wisp648 supply the power to your project. These wires need to be connected to the power rails of the breadboard.

A number of parts, such as LEDs, have a positive and negative side.[11] If you look carefully inside the LED you'll see that one of the pins appears to be wider. This is the negative side and needs to be connected to the lower voltage. Usually this wire is also

---

[11] The actual name for the positive side is the anode, and the negative side is the cathode.

shorter than the other side. In the schematic, the symbol of an LED looks like an arrow. An easy rule to remember is that the arrow always points to the negative side.

Figure 22 The positive and negative sides of a LED

Check all parts and connections twice before turning on the power. Then switch the power off again.

3.2 The software

The program is written using an editor: JALedit. Of course any other editor could be used, but this one has special features that make is very well suited for use with JAL. Once the editor has started up you'll see an empty page; this is where you'll enter the program.

A program consists of a series of commands. These commands tell the 16F877A what it needs to do. The commands are processed starting at the top of the page and going down (some commands will cause it to jump to different parts). The JAL programming language is not very particular about how you space the commands over the page. But you need to be able to understand it yourself, so we'll make a few rules for ourselves:

1. Each line will have only one command.
2. Short comments will be used to explain what the program does.
3. At loops and conditional commands we indent.

The meaning of rule three will become clear later. To make the program easier to read JALedit will color-code everything you enter. This has no impact on the functionality of the program.

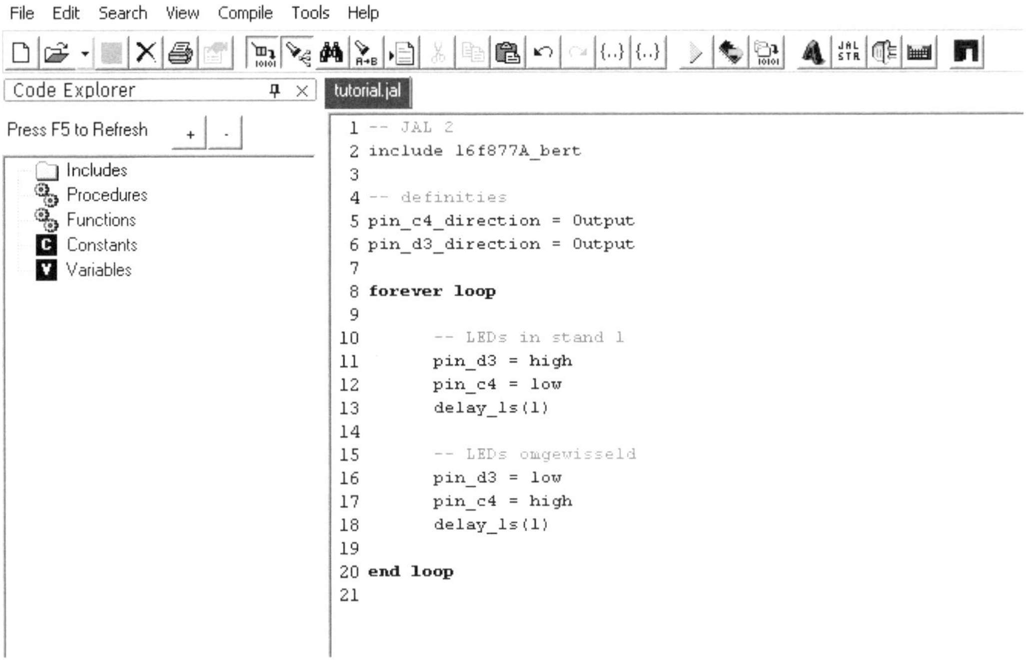

```
File  Edit  Search  View  Compile  Tools  Help

Code Explorer                      tutorial.jal

Press F5 to Refresh                 1 -- JAL 2
                                    2 include 16f877A_bert
    Includes                        3
    Procedures                      4 -- definities
    Functions                       5 pin_c4_direction = Output
    Constants                       6 pin_d3_direction = Output
    Variables                       7
                                    8 forever loop
                                    9
                                   10     -- LEDs in stand 1
                                   11     pin_d3 = high
                                   12     pin_c4 = low
                                   13     delay_1s(1)
                                   14
                                   15     -- LEDs omgewisseld
                                   16     pin_d3 = low
                                   17     pin_c4 = high
                                   18     delay_1s(1)
                                   19
                                   20 end loop
                                   21
```

Figure 23 JALedit with tutorial program.

The very first step in the program is to load a library. A library contains additional commands, which usually cannot be part of the JAL language itself because they may differ for each type of microcontroller. As a programmer you don't want to know about this stuff so it's hidden away in a library. The only thing you need to do is use the library that goes with the microcontroller. Since we are writing a program for the 16F877A, the 16F877A_bert library is loaded. The commands contained in this library are explained in Section 13, but as you read through the book you will encounter most of them, with a clear explanation.

Loading the library is done with the following command. It is a good idea, by the way, to wait before entering the program until all parts have been discussed, because this doesn't necessarily takes place in the right order. At the end of this section you'll find the complete program.

> -- JAL 2.4j
> include 16f877A_bert

The standard JAL commands are recognized by the editor and color-coded. Variables and commands from a library are in blue. The first line starts with two hyphens -- to indicate a comment only meant for the user (you); the microcontroller will do nothing with it. The second line starts with *include* and then the name of the library. Everything that this library contains is inserted at this point, without you ever seeing it.

Since the digital pins on the 16F877A can be either input or output we need to define the directions of the pins that we intend to use. For this program we want them both to be outputs, because we want to switch LEDs on and off with these pins.

    pin_c4_direction = output
    pin_d3_direction = output

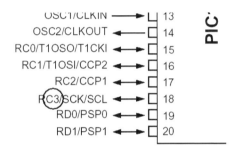

Figure 24 This is where you'll find the pin names (only c3 shown).

The pin names can be found in Figure 18. A good idea would be to look for this figure in the datasheet of the 16F877A (on the website of Microchip: http://www.microchip.com) and print it out. You can keep it handy when building projects (this is what I do).

You must indicate the direction of a pin for every pin you use. This can easily be forgotten.

Lighting an LED is done by applying power to the pin that the LED is connected to.[12] This is done by making the pin high (also called 1). We turn the power off by making the pin low (also called 0).

    pin_d3 = high
    pin_c4 = low

After these commands are executed the LED on pin d3 is on, and the LED on pin c4 is off. Since we want to make the LEDs flash, the next step has to be the reverse.

---

[12] Because the other side of the LED is connected to the ground.

```
pin_d3 = low
pin_c4 = high
```

In between a pause must be inserted. Otherwise, the flashing will be too fast to see. The pause we'll use is a one second delay:

```
delay_1s(1)
```

The number between the parenthesis specifies how many times the program should wait one second. Had we used, for example, delay_1s(3) then the program would wait 3 x 1 = 3 seconds.

With the commands used so far the LEDs will only flash once, but a real flashing light doesn't stop after a single flash. So our commands need to be repeated continuously, in fact, "forever":

```
forever loop
end loop
```

Here's what the entire program looks like:

```
-- JAL 2.4j
include 16f877A_bert

-- definitions
pin_c4_direction = output
pin_d3_direction = output

forever loop

    -- LEDs in position 1
    pin_d3 = high
    pin_c4 = low
    delay_1s(1)

    -- LEDs reversed
    pin_d3 = low
    pin_c4 = high
    delay_1s(1)

end loop
```

You see that the loop, which is repeated forever, is indented. This makes it very clear where the loop starts and where it ends. Remember, this was the third rule we made about program entry: "At loops and conditional commands we indent."

Now is a good time to enter the program into the editor and save it. In the file menu select "Save as" and enter a good name for this program, such as "tutorial". Don't enter a file extension, it will be added automatically.

It may be a good idea to use a separate directory for each project that you build. This way everything that has to do with the project is neatly grouped together.

3.3 Compiling and downloading

The first check is to see if the program has been entered correctly. Click on the button with the green triangle (compile active JAL file), or press F9. This will compile the active file.

Figure 25 Compiling the active JAL file.

You can open multiple files at the same time, but only one of them is the "active" file. For this file the tab at the top is blue. If you followed the instructions you only have one file open - the one you just entered - so this is automatically the active file.

A small window will briefly appear, and then in the "compile results" window at the bottom a message is displayed. Hopefully it starts with:

> jal 2.4j (compiled Mar 12 2009) [13]
> 0 errors, 0 warnings
> Code area: 153 of 8192 used
> Data area: 18 of 368 used
> Software stack available: 96 bytes
> Hardware stack depth 2 of 8

---

[13] If a different version of JAL is used this line will show a different number and date. This has no impact on the results. For some projects in this book version 2.4j is required.

No errors, by the way, doesn't mean your program will do what you want, or even work. It simply means you haven't made any syntax errors. Let's assume, for example, that you accidentally entered *ouput* instead of *output* you would immediately get an error message like this:

> jal 2.4j (compiled Mar 12 2009)
> [Error] (tutorial.jal) [Line 6] "ouput" not defined
> [Error] (tutorial.jal) [Line 6] '=' expected (got 'forever')
> 2 errors, 0 warnings

The compiler reports that the error has been made on line 6: "ouput" not defined. Because of this error the compiler is confused and doesn't understand what you mean by line 6 and consequently loses track of program flow. It therefor seems that you have made two mistakes. The best way to debug a program is to solve the first error and then recompile. You simply keep repeating that until there are no more errors left.

Assuming you have made no errors the program can be downloaded to the microcontroller. The hardware discussed in the previous section needs to be completed and connected to the programmer and power supply. Switch on the power[14] and on the PC click on the button with the integrated circuit and the green arrow (compile + program), or press Ctrl-F9.

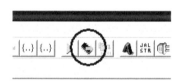

Figure 26 Compile + Program

The compile window appears again, but this time it is followed by a window with a black background. This second window belongs to the software of your programmer, in this case the Xwisp software.

---

[14] Note: these instructions only apply if you use the recommended programmer, Wisp648. If you use another programmer you need to modify the settings of JALedit to suit that programmer. Wisp648 is an in-circuit programmer. If yours isn't, then the microcontroller must be removed from the breadboard (with the power switched off) and then follow the instructions that came with your programmer.

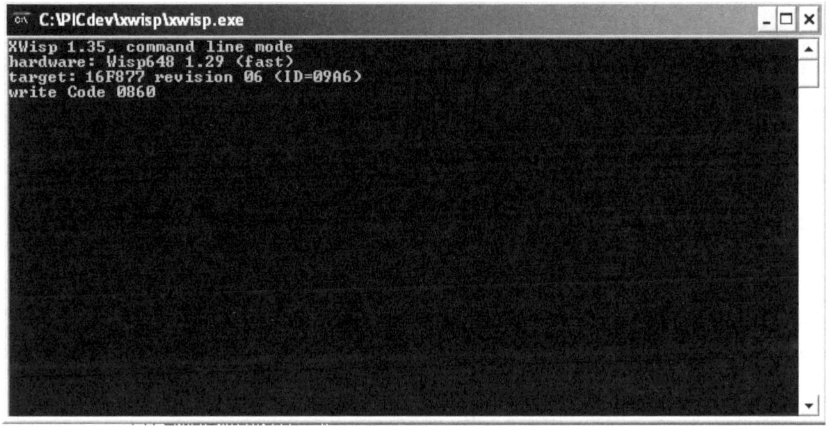

Figure 27 Programmer window.

You can observe the progress of the download in the black window. After downloading the program into the microcontroller the program is automatically checked to confirm that downloading was successful.

In the meantime check to see which version of the programmer firmware your Wisp has - this is shown in the black window. In order to use all microcontrollers used in this book you will need firmware version 1.28 or newer. If you have an older version you can perform a free upgrade as per the instructions in section 11.5 once you are done with the tutorial. This also applies to owners of the predecessor of the Wisp648: the Wisp628.

If everything is OK the LEDs will start flashing and the window will close.

3.4 Debugging

If everything went well you can skip this section and go straight to section 3.5. If it didn't go as planned look for your symptom in the table below and follow the instructions.

symptom	solution
The window with the black background doesn't come up.	1. There are errors in the program. Check the window at the bottom (compile results) for error messages. 2. You didn't install the downloaded software package correctly. Some software is missing or cannot be found, particularly the programmer software Xwisp. Carefully read the instructions of the download package and follow them exactly. 3. You have installed the software to a different location than instructed. In JALedit go to "environment options" in the "compile" menu, select the "programmer" tab and enter the correct location. Note that <u>spaces are not permitted</u> in directory or file names.
In the black window an error message appears and the window remains open. You can close the window by pressing the enter key once you have solved the problem to try again.	1. Read the error message and solve the problem. Usually the error messages are quite clear. Otherwise one of the next suggestions may solve your problem: 2. You have connected the programmer to another comport than port 1. Go to "environment options" in the "compile" menu, select the "programmer" tab and change the port to the comport number that you are using. So if you use com 4 this line would become " port com4 go %F". Note that there is no space between COM and the number. 3. The programmer is not connected or not connected correctly. Switch off the power and connect the programmer correctly (all wires!). 4. The power is off. Switch it on and retry. If you use a breadboard make sure the top and bottom power rails (and segmented rails) are connected. 5. Another program is using the same serial port as the programmer, possibly a terminal program. Close the program and try again. You may need to restart the PC if the program doesn't release the port. 6. Your power supply isn't stabile enough. Make sure the power supply of the Wisp648 has a high enough voltage (more than 9 volts) and can supply enough current (more than 500 mA). Also don't forget the 0.1 uF capacitors on then edges of the breadboard. 7. If you use your own power supply (for example because you don't own a Wisp648) make sure the voltage is stable and exactly 5 volts.

symptom	solution
The program is downloaded but the LEDs aren't flashing.	1. The LEDs are connected backwards. Insert them in the opposite direction. 2. The LEDs are connected to the incorrect pins. Switch off the power and check all connections. 3. The LEDs do flash, but extremely fast. Check if both delay statements have been entered into the program.

The program and the hardware schematic contain no errors. If despite the instructions above the program still won't work check everything again and again. Perhaps it's best to call it a day and check again tomorrow. Sometimes a good night's sleeps does wonders.

If you're still convinced that you've done everything right then meet us in the JAL usergroup at Yahoo (see http://www.boekinfo.tk for the correct address) and post your question. Note that this is an international group with users all over the world, so the mandatory language is English.

3.5 Done!

Congratulations. You've just made your first microcontroller program! In the upcoming chapters it is assumed that you will remember what you have learned in this tutorial project. In those projects you will just be shown a schematic and a breadboard photograph, and we will assume you've built it, including the power rail connectors, and the capacitors on the corners. We'll also assume that you can identity resistors and capacitors using the handy programs in the download package, so that you are capable of figuring out what the different value resistors look like.

The program itself will be shown and we will assume you've entered it into JALedit, connected the programmer, and downloaded the program into the microcontroller. Use this tutorial as a reference if you forget how to do any of these steps.

3.6 Other results

Apart from compiling and downloading, a few additional files have appeared in the program directory on your PC with the extensions of .asm and .hex. It's not necessary to know what these are for, so if you want you can skip this section and continue with the projects in Chapter 4.

The hex file

This is the file that is actually sent to the programmer to load into the microcontroller. It contains the compiled program in hexcode. Here is a fragment of it:

:10001000840AA00B07280800A0018313263084005F
:100020000330042083132B308400063004201E2864
:1000300027088700080028088000800831 2A30109
:100040008B014528AF01B0013C2883120313F23025
:10005000A0000A30A1002830A2000A128A11A20BC7

The asm (assembler) file

This is an extra file generated so that you can use your program with tools that were specifically designed to use assembler, such as the free editor/simulator offered by the manufacturer Microchip.[15]

If you look at this file you'll immediately understand why JAL is such a popular language. Each line of JAL code is shown with the assembler translation below it.

```
; 13    pin_d3 = high
                datalo_clr v__port_d_shadow
                bsf    v__port_d_shadow, 3
                call   l__port_d_flush
; 14    pin_c4 = low
                datalo_clr v__port_c_shadow 83
                bcf    v__port_c_shadow, 4
                call   l__port_c_flush
; 15    delay_1s(1)
                movlw  1
                call   l_delay_1s
```

---

[15] This program is called MPLAB® and can be downloaded for free from http://www.microchip.com. Search for MPLAB® EDI. There are paid versions as well, but the free version will do just fine.

4 Relay

In this chapter we will discus a series of projects which allow you to handle higher voltages and currents than the microcontroller itself could handle. We will use a relay for this. A relay is basically a switch that can be moved with an electromagnet. If you apply power to the coil of the electromagnet the switch will move to the other position. This subject is not very complicated and thus very suitable to discuss right after the tutorial project. It is probably a good idea to read - and try - a few of these projects before venturing to the more advanced subjects.

4.1 Auto disengaging PIC

In this project we will use a relay to let the microcontroller switch itself off.

Technical background

As relay we use a Meeder reedrelay type DIP05-1A72-12L. Most electronic components have a datasheet. This is a booklet that contains all technical specifications of that particular component. You need these data because otherwise you will not know how to connect and use the components. The relay for example has eight pins, and you need to know what these are for. You also want to know the voltage the relay requires, and how much current it uses. That is important because you need to know if you can connect the relay directly to the microcontroller without damaging it.

According to the datasheet 05 in the type number refers to the voltage, 1A refers to the contact layout, 72 for the switching model, 12 for the pin layout and L for the options. Unfortunately pin layout 12 is not listed in the manufacturers datasheet. The relay fits in a 14 pin IC socket with the pins on locations 1, 2, 6, 7, 8, 9, 13 and 14. After some experimenting it turns out that 2 and 6 belong to the coil, 1 and 14 are one side of the switch and 7, 8 and 13 are the other side. The other pins are irrelevant in this project.

The voltage is 5 volts, and that is very convenient because that is also the voltage of the microcontroller. The question is how much current the relay uses. These are the technical data:

    Coil: 5 V DC
               500 ohm
               50 mW

Using Ohm's law (voltage = current * resistance) we can calculate the current.

$$V = I * R$$

Where V is the voltage (in volts), I the current (in amperes) and R the resistance (in ohm). You can mathematically re-write the formula and enter the known data:

$$I = V / R = 5 / 500 = 0.01 \text{ ampere, or } 10 \text{ mA}$$

The 16f877A microcontroller that we use in this project can handle 25 mA per pin, and 200 mA for the entire port, so this current is not a problem. We will be using Ohm's law quite often in this book.

For the relay contacts the datasheet states:

Contact: 200 V DC or AC peak
1 A
15 W

The maximum current that the relay can handle is 1 ampere That is more than enough to switch a microcontroller on and off. The maximum voltage by the way is 200 volts, depending on where you live this may be enough to switch mains - in 110 volts mains regions - or just too little - in 240 volts mains regions.

The relay contains a coil. The current through a coil is persistent. That means that if the power is disconnected the current keeps flowing for a very short moment. This is caused by the collapse of the magnetic field. As a result the voltage over the coil reverses and may become very high. This voltage spike can seriously damage the microcontroller. The solution is to use a protection diode over the coils of the relay. This diode is mounted "backwards" so that it doesn't interfere with the normal operation but solely takes care of the reversed voltage spike. For this purpose we use a 1N4007 which has a maximum reverse voltage of 1000 volts (in our circuit it only gets 5 volts) and can handle a current of 1 A with a maximum surge current of 30 A. More than enough in our application.

Hardware

We can use any digital pin we like, so let's use pin 20 (d1) because it is located conveniently close to the relay on the breadboard. Because you can't see from the outside if the relay is engaged or not we will add a LED.

In the tutorial project you have seen how you should connect a LED. The 330 ohm resistor is to protect the LED against a too high current. A normal LED has a voltage drop of about 0.7 volts[16]. Than means that the resistor carries 5 - 0.7 is 4.3 volts. Using Ohm's

---

[16] Check the datasheet of your LED. Some LEDs are low current and are designed for 2 mA.

law you can calculate that the LED pulls 13 mA. That is low enough to cause no damage to the microcontroller.

As discussed in the tutorial project several pins must always be connected. The same applies of course in this, and all the future projects. So after this project we will not show this list anymore.

pin	name	description
1	MCLR/Vpp	Master clear pin and programmer voltage input.
11	V_{DD}	Power supply +5 volts[17]
12	V_{SS}	Power supply, 0 volts or ground
13	OSC1/CLKIN	Connection for the crystal[18]
14	OSC2/CLKOUT	Same
31	V_{SS}	Power supply, 0 volts or ground
32	V_{DD}	Power supply +5 volts

Table 6. Mandatory pins on the 16f877A microcontroller.

And of course we also need to connect the relay (and LED):

pin	description
20	relay

---

[17] All power supply pins must be connected (11,12,31,32).
[18] Technically a crystal is not mandatory because the 16f877A can run without one. However the speed will be much lower (4 MHz instead of 20 MHz) so the default library 16f877A_bert assumes we use one.

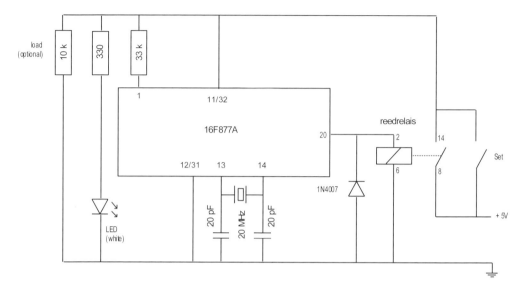

Figure 28. Schematic of the auto disengaging microcontroller.

Use the schematic to build this project on a breadboard. As you see the microcontroller is drawn as a rectangle, and only the pins that are in use are shown. This makes the schematic easier to read. Pins that are not in the schematic are not in use, and do not need to be connected to anything. Beware of the placement of the diode: the little line on it should be at the side of pin 20 - so it is mounted in reverse.

With all projects you will also see a picture of the project on a breadboard. This can be very helpful because you can see how the components are connected in real life. These pictures were taken from actual working projects so they cannot contain any mistakes. Normally only the section of the breadboard is shown that contains actual project components. In this project you can see the entire project, including the power rail connector wires and the capacitors on the corners. These components must always be used as discussed in section 2.1 So even when you don't see them they must still be placed on the breadboard.

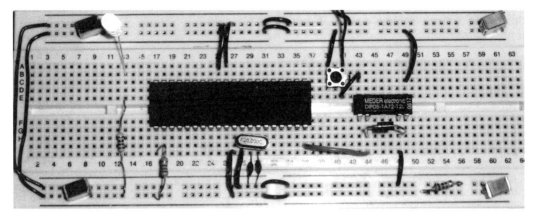

Figure 29. Picture of the project including power rail wires and corner capacitors.

Software

By placing the microcontroller behind the relay it can switch itself off by removing the power from pin 20. But in order to do that it must be switched on first. You can do that by pressing switch Set. This will route the power to the microcontroller bypassing the relay, which will engage the microcontroller (and the LED).

You have seen the command pin_d1_direction = output before. It means that pin d1 can be made high or low by the microcontroller. As soon as the program starts pin d1 is made high, so the relay will engage. You can now let go of the Set button. After 10 seconds the microcontroller will make pin d1 low again thus releasing - or disengaging - the relay which causes the microcontroller to go off, just as the LED.

```
-- JAL 2.4i
include 16F877A_bert

-- pins
pin_d1_direction = output

-- hold relay
pin_d1 = high
delay_1s(10)

-- release relay and switch microcontroller off
pin_d1 = low
```

It takes quite a while before the project is completely reset because after the LED is off the remaining power has nowhere to go. In order to be able to reset the project quickly a dummy load has been added. This 10 k resistor will "waste" residual power. In the picture it is more or less hidden in the bottom right corner, over the power rail.

To get the program into the microcontroller you need to connect the Wisp648 programmer as shown in the next table. These connections are always the same for this microcontroller, so this table will not be repeated, unless a new microcontroller is introduced.

_____ 16f877A _____	
color	pin number
yellow	pin 1
blue	pin 40
green	pin 39
white	pin 36
red	+ 5 volts
black	0
jumper	no

Table 7. Wisp programmer connections to the 16f877A.

If you have any doubt in projects where the pin connections for the Wisp are not shown then do not gamble because you can permanently damage the microcontroller if you connect it incorrectly. Section 12.2 contains list of the pin connections for all supported microcontrollers used in this book.

Operation:

1. Switch the power on.
2. Shortly press the button - the LED will go on.
3. Release the button - the LED stays on.
4. After about 10 seconds the LED will go off.

4.2 Bistable relay

In the previous project the relay is normally in one position - the so called rest position - unless you supply power to it. In that case it moves over to the other position. As soon as the power is shut off the relay goes back to the original position. This is how a normal relay work. In this project we will use a special kind of relay: the bistable relay. This relay has two rest positions. That means that you only need to supply power to flip the relay from one position to the other. After that you can cut the power and the relay will remain in that last position.

Technical background

The bistable relay RAL 5W-K manufactured by Fujitsu needs power only to flip its position. Once that has happened you can leave the power on, or cut it, whatever is more convenient. Switching the relay back to the original position again can be done by reversing the voltage over the coil. That means that you cannot use a protection diode, because a diode can only work in one direction.

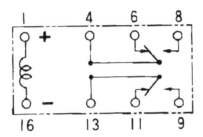

Figure 30. Pin layout of the RAL 5W-K bistable relay.

Since we do want to protect our microcontroller we need an additional chip, the TC4427A mosfet driver. This chip can handle higher voltages and currents than the microcontroller, and it can handle spikes much better. This chip is meant to drive mosfets but it can also be used for other things such as a relay or small motor.

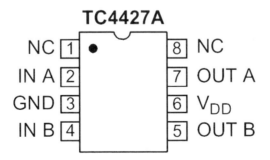

Figure 31. Pin layout of the TC4427A mosfet driver.

These are the parameters of the bistable relay:

    Coil: 5 V DC
                    167 ohm
                    150 mW

The current consumption of the relay at 5 volts is therefor $I = V/R = 5/167 = 29.9$ mA. The TC4427A mosfet driver can handle a peak current of 1.5 A and max 730 mW continuously (that is the maximum power dissipation). The relay needs 150 mW peak, so even if it would work continuously it still wouldn't be a problem. In real life the relay only uses power when it flips, which takes about 6 nS. That means in our calculations we are allowed to use the peak current which is more than enough also.

Once the relay has flipped the coil current drops to 14 mA, which is in fact wasted energy. For that reason the microcontroller will switch the current off relatively quickly.

    Contacts: 250 V AC
                          2 A
                          60 VA, 24 W

The contacts are heavy enough for our little LED. If you want to use this relay to switch 240 volts mains the situation becomes a bit more complicated. The voltage is no problem, but you need to consider the maximum values 2A, 60 VA and 24 W. The one that leads to the lowest current applies.

        60 VA results in 0.25 A as maximum.
        24 W results in 0.32 A as maximum[19].

That means the maximum is 0.25 A or a lightbulb of for example 15 W. And even that is questionable because when you switch a light bulb on the initial current can easily be 15 times the normal current. Fortunately we do not have to worry about all this for our simple LED.

Hardware

The relay fits in a 16 pin IC socket and follows that numbering, though not in the order than an IC would be numbered. The LEDs in the schematic are white and blue, but you can use any color you like.

---

[19] The formula is $P = V * I$, where P is the power in Watt.

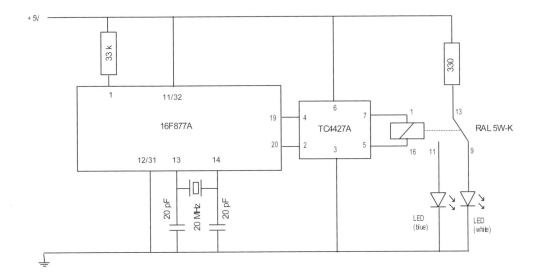

Figure 32. Schematic of the bistable relay with mosfet driver.

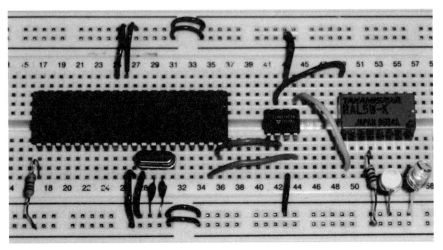

Figure 33. The project on a breadboard.

Software

According to the datasheet the time required for flipping the relay is 6 nS. In the program we use 10 ms, that should be more than enough.

When you use a variable in your program you need to indicate what the name of the variable is and what type it should be. This allows the microcontroller to reserve room in its memory.

    var bit position=0

Instead of a variable you can also use a constant. The difference is that a variable can change while the program runs, and a constant cannot.[20] In our case the position changes, so we need a variable.

You see that we immediately give position a value, namely 0. This is called initializing. PC programmers are used to new variables always being zero. In a microcontroller this is not the case! A new variable can have any value. Beware of this because it can be a very difficult to trace cause of bugs.

The position variable only has two positions in this project, namely 0 and 1. The next table shows all variable types that JAL can handle. This shows that our variable position can be defined as a bit. In general it is best to make variables as small as possible in order not to waste any room in memory and calculation time. The optimum type for a microcontroller is byte because that is the exact size of a memory location.

name	description	range
bit	1 bit unsigned boolean	0 or 1
byte	8 bit unsigned	0 to[21] 255
sbyte	8 bit signed	-128 to 127
word	16 bit unsigned	0 to 65.535
sword	16 bit signed	-32.768 to 32.767
dword	32 bit unsigned	0 to 4.294.967.296
sdword	32 bit signed	-2.147.483.648 to 2.147.483.647

Table 8. Different types of variables and their ranges.

We will introduce a new JAL command: the "if.. then.." clause. Based on a certain condition a command can be executed. Or more precisely: if that condition is true, a command will be executed.

---

[20] A constant is stored in program memory. There is more room there, so it is advisable to use constants wherever possible.
[21] "To" in this context means "up to and including".

```
if a > 3 then
    b = b +1
end if
```

In this example b will be incremented by 1 if a is larger than 3. If a is not larger than 3 then nothing happens. You can have more than one command in this clause, as long as you terminate with end if.

```
if a > 3 then
    b = b +1
    c = 4
end if
```

You can also add commands that will be executed if the condition is false, by using the "else" keyword.

```
if condition then
    do this if the condition is true
else
    do this if the condition is false
end if
```

The appendix contains list of possible operators - ">" is an operator - which you can use. Note that "=" is not an operator but "==" is, because "=" is a calculation and "==" is a question.

And this is the complete program:

```
-- JAL 2.4i
include 16F877A_bert

-- pins
pin_d0_direction = output
pin_d1_direction = output

-- variable
var bit position=0

forever loop

    -- switch relay
    if position == 0 then
```

```
    pin_d0 = low
    pin_d1 = high
    position=1
  else
    pin_d0 = high
    pin_d1 = low
    position=0
  end if
  -- wait for the relay to actually switch
  delay_10ms(1)

  -- remove power from relay
  pin_d0 = low
  pin_d1 = low
  delay_1s(1)

end loop
```

4.3 Flashing light (8 volts)

In this project we will use a solid state relay - a relay without moving parts - to flash an 8 volts AC light bulb.

Technical background

The special thing about a solid state relay is that is has no moving parts, so it is less susceptible to wear. Another advantage is that this type of relay is usually small. In this project we use the 39MF22 manufactured by Sharp. Internally this relay consists of a LED placed opposite of a light sensitive switch. If the LED is on then the light sensitive switch will conduct electricity. The switching action will happen exactly when the alternating voltage is zero - the zero cross circuitry. For this project that is irrelevant but in the next project this is important.

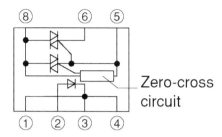

Figure 34. Pin layout of the 39MF22 solid state relay.

The pin layout looks more complicated than it is. In this project we only need pin 2 and 3 for the "coil" and pin 6 and 8 for the "switch contacts". You can tell by the LED-like symbol in the pin layout picture that pin 3 should be connected to the ground, and that the relay will switch when pin 2 is made high.

The LED inside the relay is not allowed to consume more than 50 mA according to the datasheet. The voltage drop over the solid state relay - called forward voltage in the datasheet - is 1.2 volts, so the voltage across the resistor that we need to use is 3.8 volts. And since $V = I * R$ that means that $R = V/I = 3.8 / 0.05 = 76$ ohm.

The microcontroller however can only supply 20 mA, so that means the resistor should be at least 152 ohm. To be on the safe side we use 330 ohm (red - red - brown) resistor just like we would for a normal LED

Hardware

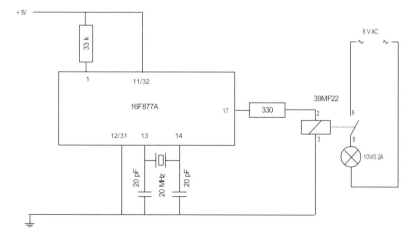

Figure 35. Schematic of the 8 volts AC flashing light.

The relay can switch a maximum of 0.9 A at a maximum of 600 volts. The light bulb we use is 0.2 A and the voltage of the transformer is 8 volts, so well within the specifications. On the picture you can see the transformer that turns mains into 8 volts AC. If you built this project make sure to cover the mains contacts, for this voltage can kill!

Figure 36. The project in action: the light is just on.

Software

This project is more about the hardware than the software, so the program is very simple. The solid state relay is switched on for on second and then off for one second.

```
-- JAL 2.4i
include 16F877A_bert

-- pins
pin_c2_direction = output

forever loop

  pin_c2 = low
  delay_1s(1)
  pin_c2 = high
  delay_1s(1)

end loop
```

4.4 Flashing light (mains, 110 or 240 volts)

In this project we will use a solid state relay - a relay without moving parts - to flash a 240 volts AC light bulb. The project will also work if your mains voltage is lower, for example 110 volts.

Technical background

In principle this project is identical to the previous one, except this time we use 240 volts mains power. This has consequences for the safety precautions. All 240 volts wires should be thicker, and it is a good idea to use a glass fuse. The relay can switch 0.9 A so an 800 mA fuse should do fine.

The second problem is that the relay can cause small rapid voltage swings on the net when switching. A snubber, consisting of a capacitor will be used to short circuit and thus eliminate these fast swings. A resistor in series with the capacitor will limit the current. As values I have selected 150 ohm (brown - green - brown) and a 10 nF (code 103) capacitor. For the resistor you can use a 1/4 W type just like the other resistors in this book, but if you plan to use this project for longer periods it is better to use a 1 W type.

Hardware

This is the complete schematic. For safety reasons the high voltage part is mounted on a separate circuit board. There is no design for this board, a small piece of perforated prototype board will do just fine.

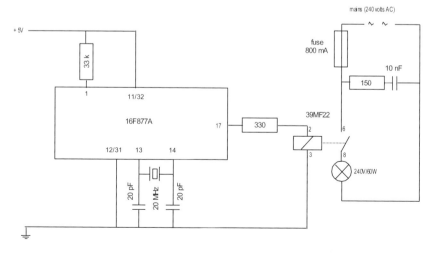

Figure 37. Schematic of the 240 volts flashing light.

> Beware that all components are connected to the mains power grid. Mains voltage can kill. So built this project in an insulating box and take all precautions necessary to prevent anyone from touching the wires. If you don't know how to do this you must not proceed.

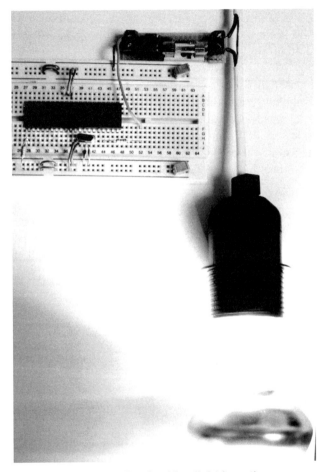

Figure 38. The flashing light in action.

Software

This project is more about the hardware than the software, so the program is very simple. The solid state relay is switched on for on second and then off for one second.

```
-- JAL 2.4i
include 16F877A_bert

-- pins
pin_c2_direction = output

forever loop

  pin_c2 = low
  delay_1s(1)
  pin_c2 = high
  delay_1s(1)

end loop
```

5 Alternating current

Alternating current means that one wire - the phase - is alternatingly positive and negative with respect to the other wire - the zero or ground. In the previous chapter we used a relay to switch AC. In this chapter we will use a triac. The advantage of a triac is that you can control the output voltage. Another advantage is that it is much faster than a relay, and since it contains no moving parts is less susceptible to wear.

Other places in this book that have AC projects, or projects that could handle AC:

Project	Name
4.2	Bistable relay
4.3	Flashing light (8 volts)
4.4	Flashing light (mains, 110 to 240 volts)

5.1 Switch (8 volts)

In this project we will use a triac to flash an 8 volts AC light bulb.

Technical background

A triac has three pins. Two of these pins are what you would call contacts if it was a relay. These pins are usually called MT1 and MT2, short for Main Terminal. These pins are connected to each other when a sufficiently high voltage is applied to the third pin, named G - for gate. The triac will stop conducting when the voltage on pin G is removed and when the current running from MT1 to MT2 drops below a certain value.

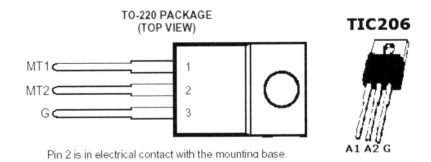

Figure 39. Pin layout of the TIC206 triac.

The voltage that needs to be applied to the gate in order for the triac to conduct is called V_{gate}, and the minimal current between MT1 and MT2 to keep the triac conducting is called the holding current, or I_{hold}. In this project we use the TIC206M triac, with the following properties according to the datasheet

	property	value
V_{max}	maximum voltage	400 V
I_{max}	maximum current	4 A
I_{hold}	minimal holding current	30 mA
V_{gate}	minimum gate voltage to conduct	2 V
$I_{gate(max)}$	maximum gate current	200 mA

Table 9. Properties of the TIC206M triac.

For the triac to function properly V_{gate} must be derived from the wire leading to MT2. And that is a problem because in this project that wire will carry AC, something the microcontroller can't handle. So we need a second component that can deliver this voltage and can be connected to the microcontroller at the same time. For this component we choose the MOC3020 optocoupler.

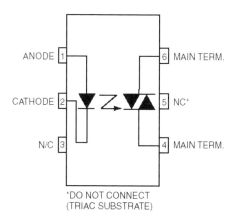

Figure 40. Pin layout of the MOC3020 optocoupler.

Internally an optocoupler looks a lot like a solid state relay. By switching the LED on - by the microcontroller - pins 4 and 6 are connected to each other. By now we know that we can connect a LED to a microcontroller if we use a 330 ohm resistor to limit the current.

The current trough the gate of the triac must be less than 200 mA, so we need a resistor for that as well. In theory this can be calculated using Ohm's law, but there is a small catch. When we speak of AC voltage then this is some sort of average value, because it keeps changing all the time. Because we are concerned about the maximum current we should be calculating with the maximum voltage, and not some average. That maximum voltage - also called peak voltage - can be calculated for a sinus-shaped AC voltage by multiplying it by the square root of two (about 1.4).

The voltage drop over the Gpin of the triac (2 volts) is not over the resistor, so this should be deducted from the voltage. That means the formula for the minimum value of the resistor is:

$$R_{min} = ((V^\sim - V_{gate}) * \text{square root } 2)/200 \text{ mA}$$

We will use a tiny light bulb that runs on 8 volts AC. That means the minimum resistance has to be

$$((8-2) * \text{square root } 2)/200 \text{ mA} = 42 \text{ ohm}$$

To be on the safe side - a transformer can easily deliver a higher voltage when unloaded - we select a much bigger resistor of 180 ohm. This results in a gate current of 47 mA. It is best to use a 1/2 W resistor because the power consumption is (8-2) V * 47 mA = 0.282 W.

Hardware

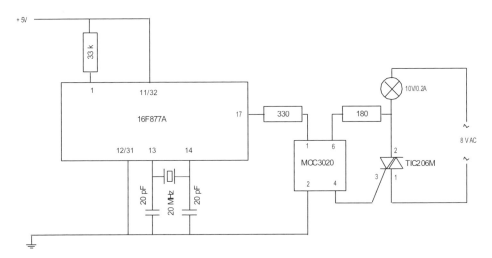

Figure 41. Schematic of the 8 volts switch with a triac and optocoupler.

The resistor for the gate has color coding brown-gray-brown. Use the schematic to built this project.

On the picture you can see the transformer that turns mains into 8 volts AC. If you built this project make sure to cover the mains contacts, for this voltage can kill!

Figure 42. The project in operation.

Software

This project is more about the hardware than the software, so the program is very simple. The triac is switched on for on second and then off for one second.

```
-- JAL 2.4i
include 16F877A_bert

-- pins
pin_c2_direction = output

forever loop

  pin_c2 = low
  delay_1s(1)
  pin_c2 = high
  delay_1s(1)

end loop
```

5.2 Zero crossing detection

Alternating current means that one wire - the phase - is alternatingly positive and negative with respect to the other wire - the zero or ground. That means that at some points the voltage is zero. In this project we will find out when that is exactly.

Technical background

Before we can start with our next project, a light dimmer, we need to solve a little problem. A microcontroller can make a pin high (+ 5 volts) or low (0 volts) but nothing in between. We can still make a dimmer, but then we need to use pulsating current. The current will be switched on, and then off again. The longer the current stays on, the longer the light is on, so the brighter it burns. If you make these pulses very short, and you give them very frequently the light doesn't get the chance to go completely off, so you will not see that the light is actually pulsating. This technique is called pulse width modulation, or PWM. The microcontroller in this project is equipped with two pins that can make use of this functionality.

In the next Figure you can see how this technique works. In the top graph the length of the pulse is 10% of the period. The period is the time between the start of two pulses. This is called the duty cycle; that period of the cycle that actually delivers duty - or power.

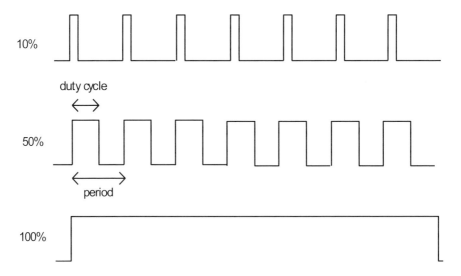

Figure 43. The effect of pulse width modulation.

In the middle graph the duty cycle is 50% and in the bottom graph the duty cycle is 100%, which means the power is on all the time and the light burns at full power.

In order to control an AC light the pulses must be given on the gate of the triac. However, when the pulse on the gate stops the triac will not go off immediately. That will only happen when the current through the main terminals drops below the holding current, so in the case of AC when the voltage is close to zero. In the previous projects, where we used the triac as a switch, this was not a problem. The mains AC has a frequency of 50 Hz[22]. That means that the voltage makes 50 waves per second, so that the current drops below the holding current 100 times per second. I'm sure you didn't notice the 1/100 second delay. But when you use PWM things start to go horribly wrong. The pulses are given very fast, and then 1/100 of a second is very noticeable.

In practice this means that when you use a triac the moment that it switches on can be determined by the microcontroller, but the moment that is switches off again is determined by the AC voltage. The frequency of the AC mains is known, so when we know when the voltage is zero we also know when the voltage will be zero the next time. This is known as the zero crossing. At a frequency of 50 Hz the voltage is zero 100 times per second. So between two zero crossings is exactly $1/100 = 10$ mS. Let's assume we want to give a 2 mS pulse. After a zero crossing we simply wait 8 mS, and then give the pulse. Two milliseconds later the next zero crossing occurs and the pulse is switched off automatically. In the next project we will discus this in some more detail. For now our assignment is to actually detect the zero crossings!

The method we use is derived from a Microchip[23] application note. One of the AC wires is connected through a very large resistor (20 M ohm at 240 volts, 10 M ohm at 8 volts) directly to pin a4. Normally this is impossible because regardless of the resistors the voltage is much too high. But pin a4 is a special pin which has an open drain - also known as an open collector. That means that on this pin (and only on this pin!) the microcontroller can handle a voltage higher than the supply voltage of 5 volts. This is what Microchip writes about this:

... is connected to the "hot" lead of the AC power line and to pin a4. The ESD protection diodes of the input structure of the GPIO allows this connection without damage (see Figure 1). When the voltage on the AC power line is positive, the protection diode from the input to VDD is forward biased, and the input buffer will see approximately $V_{DD}+0.7$ volts and the software will read the pin as high. When the voltage on the line is negative,

---

[22] In the Netherlands and most parts of continental Europe. In other countries other frequencies may be used, for example 60 Hz in the USA. If you live in such a country you will need to adapt the calculations to your own situation.
[23] DS40171A, "PICDIM Lamp Dimmer for the PIC12C508", www.microchip.com

the protection diode from VSS to the input pin is forward biased, and the input buffer sees approximately V_{SS} - 0.7 volts and the software will read the pin as low. By polling a4 for a change in state, the software can detect a zero crossing.....

The hardware

This proves that the next schematic is possible, but also that the +5 volts and the 0 wire are connected with the AC source. There is a very large resistor in between but when you are using high voltages it is still dangerous.

In the schematic you see a 10 M (brown - black - blue) resistor that connects the 8 volts AC to the microcontroller. Pin 30 is connected to the software oscilloscope WinOscillo as discussed in section 2.2 so we can see the program in action. Do remember to use the special hardware to connect the oscilloscope to your PC. Because the AC is connected to the microcontroller, which is connected to the PC it is wise to do this experiment with 8 volts as shown in the schematic and never with mains.

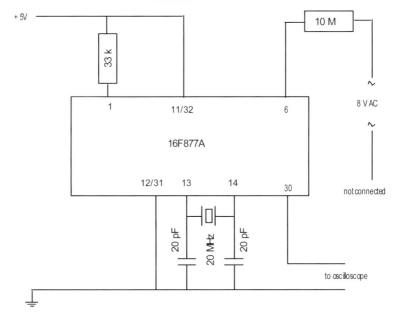

Figure 44. AC zero crossing detection.

Figure 45. Zero crossing detector on a breadboard.

In the picture of the breadboard the right wire from the transformer is not connected to anything.

The software

In this program we will wait for pin a4 to go low - a zero crossing from positive to negative - with a while loop:

    while pin_a4 loop end loop

You will probably find it easier to read when it is not on a single line but you will find this one-liner quite often in software.

    while pin_a4 loop
    end loop

This means that while pin a4 is true the loop must be executed. In the loop nothing happens, so basically it means: wait here until pin a4 is no longer true. For a microcontroller true is the same as high, or 1. So this construction also means: wait here until pin a4 is no longer high.

pin voltage	description
+ 5 volts	1, true, high
0	0, false (not true), low

As soon as the pin is low the program will continue. It will now give a pulse on pin d7 (the oscilloscope is attached to that pin) of 50 uS. The command delay_10us(5) means wait 10 uS and do that 5 times, so 50 uS in total. Then the program waits for the next zero crossing. This time the voltage will swing the other way so from low to high. That means we now have to wait until the pin is high again, or until it is no longer not true:

while !pin_a4 loop end loop

The symbol for not is the exclamation mark !. This symbol is called an operator, much like > and + are operators. You will find a complete overview of all operators in the appendix.

```
-- JAL 2.4i
include 16F877A_bert

-- pins
pin_d7_direction = output
pin_a4_direction = input

forever loop

   -- wait for the pin to go low
   while pin_a4 loop end loop
   -- and give a 50 uS pulse on d7
   pin_d7 = 1
   delay_10us(5)
   pin_d7 = 0

   -- wait for the pin to go high
   while !pin_a4 loop end loop
   pin_d7 = 1
   delay_10us(5)
   pin_d7 = 0

end loop
```

On the WinOscillo display you can see that two zero crossings together have a length of 20 mS. This matches a zero crossing frequency of 100 Hz, which is correct because the mains frequency is 50 Hz and each AC wave contains 2 zero crossings.

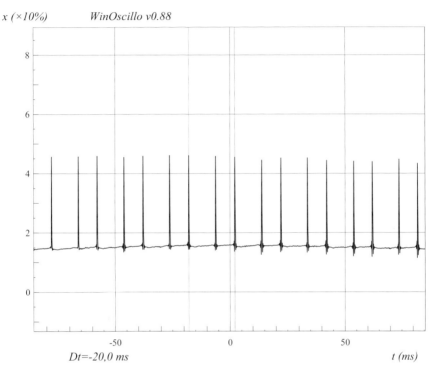

Figure 46. The length of a wave (two zero crossings) is 20 ms.

5.3 Light dimmer (8 volts)

In this project we will use a triac to make an 8 volts light dimmer.

Technical background

In the case of AC, alternating current, it is impossible to use PWM. The frequency of the AC power will never be exactly synchronous with the PWM frequency. The result is that each time a different part of the AC wave is cut off. That results in a light that fluctuates in brightness. For that reason we will use the zero crossings to keep the AC frequency and the frequency of our phase trimming synchronous.

This principle is simple. After a zero crossing it will take about 10 mS for the next one to arrive[24]. At each zero crossing the triac will switch off automatically, so all we need to do it switch it on at the right moment. The longer we wait, the less power will go to the light. In the next Figure you can see this effect. In the top graph we waited longer than in the bottom graph, so less power (the hatched area) is send to the light. This means it will burn less bright. It looks as if we are trimming part of the wave away, and since we do this from the beginning of the wave this is called "leading edge trimming".

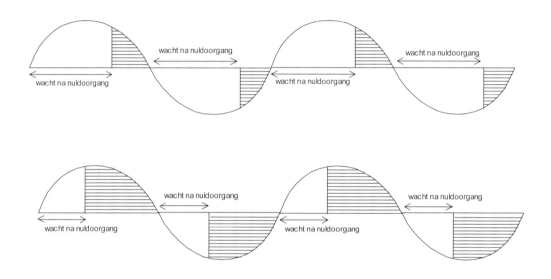

Figure 47. Leading edge trimming and zero crossing.

The frequency of the wave is 50 Hz which means that the zero crossings have a frequency of 100 Hz, for each wave has two zero crossings. The maximum waiting time at a frequency of 100 Hz is 1/100 S = 10 mS. It is convenient to divide this time up in 255 steps - we will discuss later why - so one step is 10 mS / 255 = 39 uS. In reality the microcontroller also needs to execute the rest of the program and if we would wait 39 uS it would not have time to do so, and we would end up in the next wave. So we need to use a slightly shorter time, namely 37 uS. There is no delay with this exact value - in the appendix you will find a complete overview - so we will make one ourselves.

We can do this by making a procedure. A procedure is a homemade JAL command. The libraries that you have included by using include 16f877A_bert are filled with these

---

[24] In the Netherlands and most parts of continental Europe. I other countries other frequencies may be used, for example 60 Hz in the USA. If you live in such a country you will need to adapt the calculations to your own situation.

additional commands. Making a procedure is very simple. We will start by selecting an appropriate name, in this case for example delay_37uS. We start with the shell of the procedure:

> procedure delay_37uS is
> end procedure

This states the name of the procedure, where it starts, and where it ends. Now we need to figure out which data this procedure needs to get from the normal part of our program. In our case we want to wait m37 uS multiple times (up to 255 times), so we introduce a variable called N which will be between 1 and 255 (zero would be pointless) so a byte will do fine. We want to give the value of N to the procedure, so from the main program - the outside - into the procedure - going in. So we use the definition "byte in".

> procedure delay_37uS (byte in N) is
> end procedure

This defines N as a byte, but it is only known inside of this procedure. We don't need to get any data out of the procedure (otherwise we would have used byte out) so now we can write then content of the procedure. This is no different than writing a normal JAL program. JAL has a built in delay of 1 uS called "_usec_delay()". We can use this command to wait 37 uS. So all we need to do now is repeat that delay N times. For this we can use the "for...loop" command. This works the same as the forever loop command, except now the loop is not repeated forever but a fixed number of times. Putting it all together gives us this procedure:

> -- special delay routine
> procedure delay_37uS (byte in N) is
>   for N loop
>   _usec_delay(37)
>   end loop
> end procedure

If for example you want to wait 15 times 37 uS (555 uS) then you can call this procedure like this
> delay_37uS(15)

This procedure must be listed in the program before it is used for the first time. JAL reads the program from top to bottom, so if you enter the call first and the procedure later then JAL has no idea what you are referring to and generates an error

The 16f877A microcontroller has pins on the A and E port that are capable of converting an analog signal to a digital value. In this project we will make use of this functionality by connecting a variable resistor to pin 2. Such a variable resistor is also called potentiometer (colloquially known as a pot or potmeter). Of this potmeter one side is connected to the ground, and the other to the +5 volts. The wiper - the middle contact - can now be varied between 0 and +5 volts simply by turning the axle. For analog to digital conversion (A/D) you can use the command adc_read_low_res which will give you a digital representation of the analog value in a range from 0 (0 volts) to 255 (+5 volts).

> resist = adc_read_low_res(0)

The number between brackets is not the pin number but the number of the analog to digital converter in the microcontroller. The abbreviation for this unit is ADC or A/D, but in the datasheet it is called AN (from ANalog) channel. The next table shows which AN channel is connected to which pin. Note that pin a4 is skipped.

number	name	AN
2	a0	0
3	a1	1
4	a2	2
5	a3	3
7	a5	4
8	e0	5
9	e1	6
10	e2	7

Table 10. Relation between the pin number and AN unit.

AN converter 0 is connected to pin a2 so the software matches the connection of the potmeter. The result of this A/D conversion is:

$$\frac{V_{measured}}{V_{reference}} \times 255$$

In our case the reference voltage is 5 volts, so our range of digital values is 0 to 255. Now is it clear why it was convenient a few pages ago to divide the time between two zero crossings in 255 steps. For now we can have the position of the potmeter directly control the waiting time.

```
resist = adc_read_low_res(0)
delay_37us(resist)
```

There are more A/D related commands available to you, which are all listed in the appendix. These are the most important ones:

command	description
ADC_on	Switch A/D on (for all channels).
ADC_off	Switch A/D off (all pins become digital).
ADC_read(chan)	Read the analog value on a channel into a word, max 1023. Channel is the AN number in the table.
ADC_read_bytes(chan, Hbyte, Lbyte)	Read the analog value on a channel into two separate bytes.
ADC_read_low_res (chan)	Read the analog value on a channel into one byte, with low resolution (max value: 255).

Table 11. Some A/D commands used in the projects.

Hardware

The optocoupler and triac are connected as normal. The potmeter is connected to pin 2. The longer the waiting time the less bright the light will be. For that reason the leftmost pin of the potmeter is connected to the +5 volts, and the rightmost pin to the ground. If you turn the knob right the light burns brighter. You might think the potmeter is actually using power since the end pins are connected to +5 and the ground, and you are correct. The current that flows through the potmeter is $I = V / R = 5 / 1000 = 5$ mA.

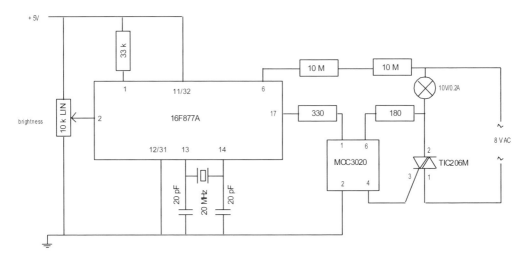

Figure 48. Light dimmer for an 8 volts AC light.

On the picture you can see the transformer that turns mains into 8 volts AC. If you built this project make sure to cover the mains contacts, for this voltage can kill!

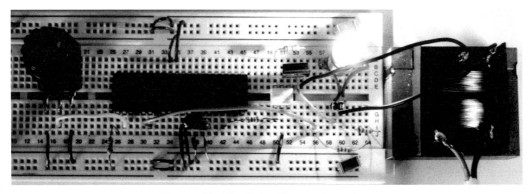

Figure 49. The light dimmer in action.

Software

The first step in this program is to determine the position of the potmeter - the knob. This position is stored as a digital value in variable called resist. Next we wait for this first zero crossing. As soon as that has been detected the gate of the triac is made low. This way we know for sure that the previous pulse was long enough. Since this happens at a zero crossing the triac switches off. Next we wait, in steps of 37 uS. Once the waiting time is finished the gate is made high again and the power to the light is thus engaged, For the remaining part of the curve the light remains on.

>   -- get position of the variable resistor
>   resist = adc_read_low_res(0)
>
>   -- wait for the pin to go low
>   while pin_a4 loop end loop
>   -- wait and the switch the triac on
>   pin_c2 = 0
>   delay_37us(resist)
>   pin_c2 = 1

The next step is to do the exact same thing at the next zero crossing, which will be in the other direction. It is not necessary to do another A/D measurement because you will not be able to spin the knob fast enough for that to be relevant. So we will only do that once in the main loop.

If you use a variable in your program you need to declare it, by stating its name and type. The variable resist will contain values from 0 to 255 so a byte will do.

>   var byte resist

Instead of a variable you can also use a constant. The difference is that a variable can change while the program runs, and a constant cannot.[25] In our case the position changes, so we need a variable.

Putting it all together gives this result::

>   -- JAL 2.4i
>   include 16F877A_bert
>
>   -- pins
>   pin_c2_direction = output
>   pin_a4_direction = input
>   pin_a0_direction = input
>
>   -- define variables
>   var byte resist

---

[25] A constant is stored in program memory. There is more room there, so it is advisable to use constants wherever possible.

```
-- special delay routine
procedure delay_37uS (byte in N) is
  for N loop
    _usec_delay(37)
  end loop
end procedure

-- main program
forever loop

  -- get position of the variable resistor
  resist = adc_read_low_res(0)

  -- wait for the pin to go low
  while pin_a4 loop end loop
  -- wait and the switch the triac on
  pin_c2 = 0
  delay_37us(resist)
  pin_c2 = 1

  -- wait for the pin to go high
  while !pin_a4 loop end loop
  -- wait and the switch the triac on
  pin_c2 = 0
  delay_37us(resist)
  pin_c2 = 1

end loop
```

5.4 Light dimmer (mains, 110 to 240 volts)

In this project we will built a light dimmer for 240 volts mains. This project will also work if your mains voltage is lower, for example 220 volts or 110 volts. It is designed for 50 Hz. If your mains has a different frequency you need to modify the calculations in the previous section and use the result here.

Technical background

This project is identical to the previous project, except this time you must use a 240 volts light bulb and mains as AC power.

Hardware

1. In this project we use lethal mains power. Build this project only if you have the knowledge to work with dangerous voltages like that.
2. You can use a breadboard only if it is certified for mains power. If it is not, or when you are in doubt, build the project on a printed circuit board.
3. Disconnect the Wisp648 programmer from the breadboard and from the PC before applying the 240 volts to the project. That means disconnect the RS232 plug from the Wisp and all wires from the breadboard. If you use the Wisp648 to power your project you can leave the red and black wires connected (but only those two!)
4. Make sure no part of the project can be touched by humans or animals. The entire triac and numerous other parts are connected directly to mains, and this voltage can kill!
5. If you use a mains certified breadboard only use a small wattage light bulb. If you want to use other light bulbs make sure to use thicker wires and mount at least the optocoupler and triac on a separate printed circuit board with wide enough copper lanes.
6. Built a permanent set-up in an insulated box and use a printed circuit board with sufficiently thick copper lanes.
7. Use a potmeter with a plastic casing and a plastic axle.

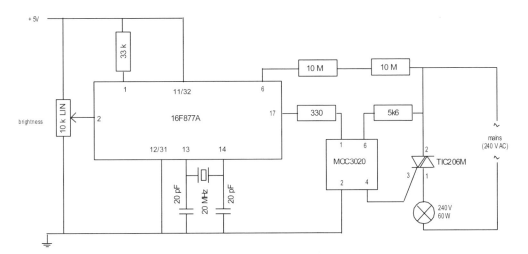

Figure 50. The mains light dimmer.

> Beware that all components are connected to the mains power grid. Mains voltage can kill. So build this project in an insulating box and take all precautions necessary to prevent anyone from touching the wires. If you don't know how to do this you must not proceed.

The setup as shown in the next Figure is for demonstration purposes only, connected directly to mains power, and thus dangerous. Make sure to read, understand and obey the safety instructions. Mains power can kill!

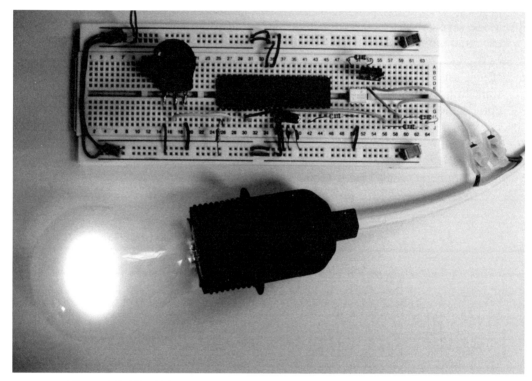

Figure 51. The light dimmer, make sure to read the safety instructions.

Software

The software is identical to the previous project.

```
-- JAL 2.4i
include 16F877A_bert
-- pins
pin_c2_direction = output
```

```
pin_a4_direction = input
pin_a0_direction = input

-- define variables
var byte resist

-- special delay routine
procedure delay_37uS (byte in N) is
   for N loop
      _usec_delay(37)
   end loop
end procedure

-- main program
forever loop

   -- get position of the variable resistor
   resist = adc_read_low_res(0)

   -- wait for the pin to go low
   while pin_a4 loop end loop
   -- wait and the switch the triac on
   pin_c2 = 0
   delay_37us(resist)
   pin_c2 = 1

   -- wait for the pin to go high
   while !pin_a4 loop end loop
   -- wait and the switch the triac on
   pin_c2 = 0
   delay_37us(resist)
   pin_c2 = 1


end loop
```

6 Make sound

In this chapter we will discuss a series of projects that make sound. Making sound usually involves great timing accuracy, and thus interrupts. In the last project of this chapter you can make your microcontroller talk.

6.1 Youth deterrent

Many people find groups of teens hanging about annoying and look for non-violent ways to keep their doorsteps empty. One solution is to play a high pitched whistle at a frequency that only teenagers can hear. The target frequency is 16 to 20 kHz. If you are a teenager yourself you can use this project to make a secret doorbell or warning signal that adults cannot hear.

Technical background.

OPTION_REG REGISTER

R/W-1	R/W-1	R/W-1	R/W-1	R/W-1	R/W-1	R/W-1	R/W-1
RBPU	INTEDG	T0CS	T0SE	PSA	PS2	PS1	PS0
bit 7							bit 0

bit 7	**RBPU**
bit 6	**INTEDG**
bit 5	**T0CS**: TMR0 Clock Source Select bit 1 = Transition on T0CKI pin 0 = Internal instruction cycle clock (CLKOUT)
bit 4	**T0SE**: TMR0 Source Edge Select bit 1 = Increment on high-to-low transition on T0CKI pin 0 = Increment on low-to-high transition on T0CKI pin
bit 3	**PSA**: Prescaler Assignment bit 1 = Prescaler is assigned to the WDT 0 = Prescaler is assigned to the Timer0 module
bit 2-0	**PS2:PS0**: Prescaler Rate Select bits

Bit Value	TMR0 Rate	WDT Rate
000	1 : 2	1 : 1
001	1 : 4	1 : 2
010	1 : 8	1 : 4
011	1 : 16	1 : 8
100	1 : 32	1 : 16
101	1 : 64	1 : 32
110	1 : 128	1 : 64
111	1 : 256	1 : 128

Figure 52. The option_reg register of the 16f877A microcontroller.

A whistle is a sound with a single frequency. That means that the loudspeaker must receive an alternating voltage in this same frequency. The easiest way to make this signal is to flip a pin from high to low and back at the right speed. Because we need to do that at a high frequency, but more importantly very accurately we will use an interrupt. That means that at regular intervals the main program is interrupted to execute a few other commands.

For this interrupt we can use one of the special timers that the microcontroller has, for example timer zero - usually written as timer0. This timer can interrupt a program to run a section of separate code. How often it will interrupt the program is controlled by the option_reg register. Registers are special places in the memory of the microcontroller that are used to store settings. There are dozens of these registers for all sorts of purposes, many of which we will discuss in this book. What each of the registers is for is described in detail in the datasheets of the microcontroller. In this section we use the 16f877A so we will use registers that this microcontroller uses. If you want to use this program on another microcontroller you may need to change the register names and perhaps the content as well.

This same register can by the way also be used to generate interrupts on port B. RBPU (port B pull-up enable) and INTEDG (interrupt edge select) are meant for this. In this project we do not need them, but we do in section 11.4 where we will discuss this in more detail.

T0CS (Timer0 clock source) indicates if you are feeding the timer internally or externally. Since a crystal is used you might think that this qualifies as external. In this case however external refers to pin T0CK1 - pin a6 - so we select internal.

The next bits (PS0, PS1, PS2 and PSA) determine how often the interrupt should occur. Time for some math. The crystal has a frequency of 20 MHz. Every <u>forth</u> pulse timer0 is incremented by one (because every instruction needs four ticks of the clock, or clock ticks). An interrupt is generated when timer0 overflows, so after 256 additions. That means the standard interrupt frequency is:

$$20{,}000{,}000 / 4 / 256 = 19{,}531 \text{ Hz}$$

That is a tad quick so a prescaler may be used to slow the interrupt rate down. This prescaler can be assigned to the WDT (watchdog timer) or the timer0 module. What that all means is shown in the following Table.

prescaler	assigned to TMR0 frequency (Hz)	assigned to WDT frequency (Hz)
0	9,766	19,531
1	4,883	9,766
2	2,441	4,883
3	1,221	2,441
4	610	1,221
5	305	610
6	153	305
7	76	153

Table 12. Interrupt frequencies of the 16f877A microcontroller.

If for example the prescaler is set to two and the interrupt is assigned to TMR0 then the interrupt frequency is 2,441 Hz. For that bit 1 must be on and the rest off, so the option_reg register would in that case be 0b_0000_0010. This is by the way the frequency of the interrupt. Sound consists of a wave, and a wave has a high part and a low part. The sound frequency is therefor half the interrupt frequency. So the sound frequency would be 1,220 Hz.

It is also possible to increase the frequency. Normally timer0 will start at zero after an overrun, but you could also make it start at a higher value. This means the overrun point is reached sooner which increases the interrupt frequency. If timer0 for example starts at 50 instead of zero then the overrun time would be (256 - 50) * 0.2 uS, so a theoretical interrupt frequency of 24,272 Hz.

When the interrupt frequency is too high the main program itself doesn't have time to run anymore so the program will crash. If timer0 would start at say 200 then there are only 56 instructions left for your entire program, including the interrupt routine itself and the interrupt overhead. So for the higher starting values the next table is more theory than practice.

	assigned to WDT, prescaler = 0	
tmr0 start	interrupt frequency (Hz)	sound frequency (Hz)
0 (do nothing)	19,531	9,766
50	24,272	12,136
100	32,051	16,023
150	47,170	23,585
200	89,286	44,643

Table 13. Impact of an increased timer0 start value.

The procedure that is executed during an interrupt looks exactly the same as a normal procedure, with a few remarks.

1. The very first line of the procedure is "pragma interrupt". This is an instruction to the compiler to turn this procedure into an interrupt procedure. You can make as many of these procedures as you want, the compiler will combine them for you.
2. Interrupt procedures may never be called by the main program itself (or by any other procedure for that matter).

An interrupt procedure has this format:

procedure [use any name] is
   pragma interrupt

   [this is where your commands go]

end procedure

The only thing left to do is to actually switch the interrupt on. This takes place using the intcon register[26]. Again we consult the datasheet.

---

[26] Some routines shouldn't be interrupted while they are in progress; for example software based communication routines or writing in flash memory. Using this register you can easily disable interrupt before you start any of these things. That does of course mean that your fixed frequency is gone.

INTCON REGISTER (ADDRESS 0Bh, 8Bh, 10Bh, 18Bh)

R/W-0	R/W-0	R/W-0	R/W-0	R/W-0	R/W-0	R/W-0	R/W-x
GIE	PEIE	T0IE	INTE	RBIE	T0IF	INTF	RBIF
bit 7							bit 0

bit 7   **GIE**: Global Interrupt Enable bit
1 = Enables all unmasked interrupts
0 = Disables all interrupts

bit 6   **PEIE**: Peripheral Interrupt Enable bit
1 = Enables all unmasked peripheral interrupts
0 = Disables all peripheral interrupts

bit 5   **T0IE**: TMR0 Overflow Interrupt Enable bit
1 = Enables the TMR0 interrupt
0 = Disables the TMR0 interrupt

bit 4   **INTE**: RB0/INT External Interrupt Enable bit
1 = Enables the RB0/INT external interrupt
0 = Disables the RB0/INT external interrupt

bit 3   **RBIE**: RB Port Change Interrupt Enable bit
1 = Enables the RB port change interrupt
0 = Disables the RB port change interrupt

bit 2   **T0IF**: TMR0 Overflow Interrupt Flag bit
1 = TMR0 register has overflowed (must be cleared in software)
0 = TMR0 register did not overflow

bit 1   **INTF**: RB0/INT External Interrupt Flag bit
1 = The RB0/INT external interrupt occurred (must be cleared in software)
0 = The RB0/INT external interrupt did not occur

bit 0   **RBIF**: RB Port Change Interrupt Flag bit
1 = At least one of the RB7:RB4 pins changed state; a mismatch condition will continue to set the bit. Reading PORTB will end the mismatch condition and allow the bit to be cleared (must be cleared in software).
0 = None of the RB7:RB4 pins have changed state

Figure 53. The intcon register of the 16f877A microcontroller.

This register is part settings, part information. For us bits 7, 5 and 2 are interesting. Bit 7 is used as a general interrupt on/off switch, and bit 5 is used to enable the timer0 interrupt itself. So:

    intcon = 0b_1010_0000

By now you may have noticed that interrupts can take place for many different reasons. At each interrupt, regardless of the reason, the interrupt routine is called. To make sure that this is "our" interrupt bit 2 can be used. If this bit (t0if) is set - has value one - at the moment that the interrupt takes place then this is a timer0 interrupt. So the procedure should check this, and clear this flag for the next time.

```
procedure Make_a_sound is
    pragma interrupt

    if t0if then

        [this is where your commands go]

            t0if = 0
    end if

    end procedure
```

The interrupt routine is simple. All we need to do is flip the pin. The easiest way to do this would be like this:

```
pin_d2 = !pin_d2
```

This means pin d2 is not pin d2. In other words: make pin d2 what it is not at this moment. So on becomes off and off becomes on. And with a slowly flashing LED this will certainly work. But in order to do this the microcontroller must first look at the pin to see what the current state is. Let's assume that the pin has just been made high, but the device that is connected to that pin has a bit of trouble following that change rapidly. The next time the microcontroller looks the pin might still be low. So the microcontroller will make it high. Again. And gone is our steady rhythm. This phenomena will occur at high(er) frequencies with devices such as capacitors and coils. A loudspeaker contains a coil so it is very likely that we will run into this problem. The solution is easy. We will use an intermediary variable - called flag - to remember what the current state of the pin is, or at least should be.

```
flag = !flag
pin_d2 = flag
```

This completes the interrupt procedure. Back to our project. The intention is to make a whistle at a high frequency that only teenagers can hear. It turns out that 16 kHz or higher is a good value. By a strange coincidence table 13 contains a correct starting value for timer0 to achieve this frequency. A WDT assigned interrupt, prescaler = 0, and timer0 start value 100 results in a sound frequency of 16,023 Hz.

We already said that there is a difference between theory and practice due to program overhead. From the next table with actual measurements it shows that a starting value of 135 gives the best result (16,2 kHz).

tmr0 start	sound frequency (Hz)	
	calculated	measured
0 (do nothing)	9,766	9.7 kHz
50	12,136	10.4 kHz
100	16,023	13.1 kHz
125	19,084	15.1 kHz
135	20,661	16.2 kHz
150	23,585	17.9 kHz
200	44,643	-

Table 14. Comparing theory and practice for timer0 start values.

Notations like 0b_0000_0010 may have surprised you. This is a very convenient way to write numbers, particularly for use with registers. This is the so called binary notation. Normally speaking humans use decimal numbers. The number 135 means one hundred and thirty-five. Actually it is just a list of three digits, but we have agreed that these digits have a different meaning, depending on their position in the row. The 5 is at the far right, and represents five. The 3 is on the second position (we start counting positions from the right) and represents thirty. The 1 is on the third position and represents one hundred. You are as it were using this table, except that computers see the first position as 0 whereas you will see it as 1.

position	3	2	1	0
multiplier	times thousand	times hundred	times ten	times on

This is called the decimal system, because the multiplication can be written in powers of ten:

position	3	2	1	0
multiplier	times 10^3	times 10^2	times 10^1	times 10^0

For computers counting with just two numbers is more convenient, 0 and 1, off and on, low and high. This is called the binary system.

position	3	2	1	0
multiplier	times 2^3	times 2^2	times 2^1	times 2^0

The binary number 101 is $1 * 2^2 + 0 * 2^1 + 1 * 2^0 = 5$ in decimal. The problem is that just by looking at 101 you cannot see if this is a binary number. This does come in handy in jokes: there are 10 types of people in this world: those who can count in binary and those who can't. To prevent this confusion we use a prefix, and for binary this prefix is 0b.

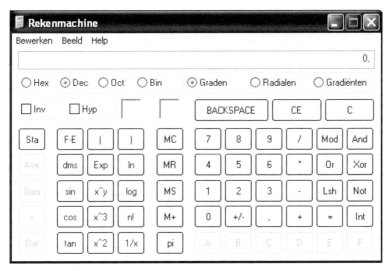

Figure 54. Windows calculator (XP SP2 Dutch).

It is very convenient if you can convert binary to digital and back by heart, but if math isn't your forte you can also use the Windows Calculator which comes for free with Windows. You must switch it to scientific mode (in Windows 7 in programmer mode) to use these features.

So this is a valid statement to use:

    intcon = 160

But it is much less clear regarding which bits are on and off than this:

    intcon = 0b_1010_0000

Hardware

The loudspeaker that I use is from and old headphone and uses about 1.54 mA, so it is a high impedance speaker. It can be connected directly to the microcontroller. If your loudspeaker draws more current you need to be careful. The coil in the speaker can cause voltage spikes that damage your microcontroller. In that case it is advisable to use a different loudspeaker, or an 100 ohm resistor in series with it.

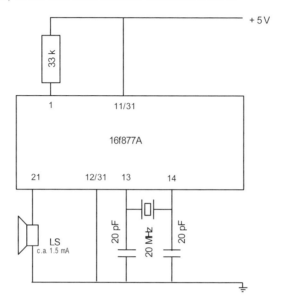

Figure 55. Schematic of the youth deterrent.

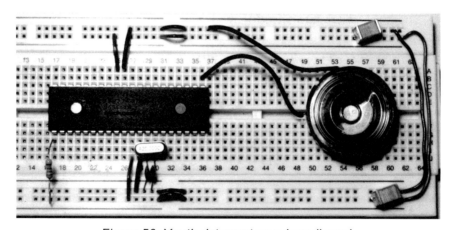

Figure 56. Youth deterrent on a breadboard.

De software

In the program you will recognize the interrupt routine and the register settings. The program doesn't have to do anything but wait for the interrupt. For that reason the main program consists of an empty forever loop. You must have this loop otherwise the program will stop, and so will the interrupt. If the program isn't running it cannot be interrupted after all.

```
--JAL 2.4i
include 16f877a_bert

-- define the speaker pin
pin_d2_direction = output

-- define the flag and set the starting value
var bit flag = 0

procedure Make_a_sound is
  pragma interrupt

  if t0if then
    -- timer0 overflow interrupt
    flag = !flag
    pin_d2 = flag

    -- set the timer0 starting time
    tmr0 = 135

    -- clear the interrupt
    t0if = 0
  end if
end procedure

-- enable timer0
option_reg = 0b_0000_1000

-- enable interrupts
intcon = 0b_1010_0000

-- nothing to do in the main loop
forever loop
end loop
```

In the download package you will find a spectrum analyzer. In order to connect this to your PC you must use the same hardware as for WinOscillo - see section 2.2. Connect the ground wire to the ground of your project and the measuring wire to pin d2. Turn the potmeter on the WinOscillo hardware to zero, switch the project on and start the spectrum analyzer. Click on the run button to start the program. Turn the knob up until you see a peak in the signal, as shown in the next Figure. Click with the mouse near the high peak to move the blue line. The small red line will now search for the strongest signal near the blue line, and this value is shown at the bottom right corner.

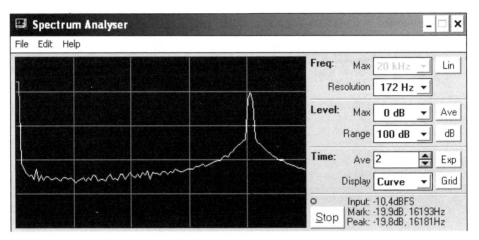

Figure 57. Spectrum of the youth deterrent

When I made this project I had no idea if it actually worked, for I couldn't hear it. The spectrum analyzer did show a strong peak at 16,2 kHz, but would it be annoying enough? So when my teenage daughter walked into my office I switched it on and I was just about to ask "Do you hear anything?" when she jumped back, covered her ears and yelled "What is that horrible noise!? "

So I guess it works just fine. Note that in some countries the use of this device on public or semi-public premises is illegal. So check your local laws before using it.

6.2 Digital to analog (D/A)

In project 5.3. we discussed how to convert an analog signal to a digital one (A/D conversion). In this project we will do the reverse. We will convert a digital signal into an analog one (D/A conversion). We will use PWM with a hardware filter. In section 9.16 we will show a different technique that uses a special IC.

Technical background

In section 5.2 we briefly discussed PWM. We saw that pulse width modulation uses pulses with a fixed frequency but a variable width. The relationship between the width of the pulses and the frequency is called duty cycle. Since we can vary this width as we please PWM basically consist of two mixed signals. A high frequency signal - the frequency of the PWM pulses themselves - and a low frequency signal - the rate of change in the width of the pulses when the user adjusts the duty cycle. If we remove the high frequency signal we are left with the low frequency one, and that will be an analog representation of the duty cycle.

So what we need is a filter that blocks high frequencies but allows the low frequencies to pass, a so called "low pass filter". Such a filter will have the following basic schematic:

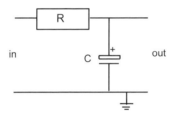

Figure 58. Principle of an RC filter.

A network such as this is called an RC network or an RC filter. R stands for resistor and C for capacitor. The capacitor can be an electrolyte - in which case the + should be on the side of the resistor - but this is not a requirement. We assume that nothing - or something with a very high resistance - is connected to the filter. That way we can assume that the current trough R is identical to the current through C and that simplifies the calculations quite a bit. The main variable of an RC network is the amount of power that will pass through at a given frequency.

$$p = g^2$$

Where p = power
g = gain

The gain is the amplification of the RC network. The gain is calculated using this formula:

$$g = \frac{1}{\sqrt{1+(wRC)^2}}$$

Where R = resistance (ohm)
       C = capacitor (F)
       w = phase angle (rad)

The phase angle is the shift in the alternating voltage when it passes through an RC filter. This shift depends on the frequency and can be calculated as follows:

$$f = \frac{w}{2 * pi}$$

This is getting rather complicated, so in the download package you will find an Excel spreadsheet to calculate RC filters. The basic question is which frequencies do we want to maintain, and which should be removed. The boundary between those two is called the cut-off frequency. The spreadsheet allows you to enter five frequencies, and of course values for R and C, and then it will calculate a power graph. It will also calculate the cut-off frequency.

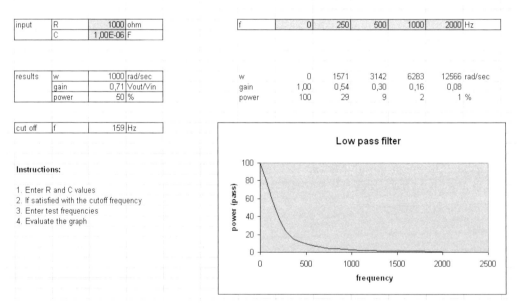

Figure 59. Excel spreadsheet, R = 1 k, C = 1 uF, cut-off frequency = 159 Hz.

As a resistor we chose a value of 1 k, and we vary the capacitor until we get a result that looks good. Based on the spreadsheet we select a capacitor of 1 uF that results in a cut-off frequency of 159 Hz.

Hardware

In this project we use a different microcontroller, namely the 18f4455. This 40 pin microcontroller has more memory that the 16f877A, a higher speed and the possibility to use USB. In section 12 you can see how you can use a different microcontroller than the one I suggest. Since this is the first time that we use the 18f4455 lets compare the properties and the Wisp648 connections.

PIC	Program (words)	RAM (bytes)	EEPROM (bytes)	I/O pins	Analog inputs	RS232	USB	I²C	SPI	PWM	CAN bus	Speed (mips)
16f877A	8192	368	256	33	8	2	0	2	2	2	0	5
18f4455	12288	2048	256	35	13	2	1	2	2	2	0	12

Table 15. Comparison between the 16f877A and 18f4455 microcontroller.

PIC	yellow	blue	green	white	red	black	jumper
16f877A	1	40	39	36	+ 5V	0	no
18f4455	1	40	39	38	+ 5V	0	no

Table 16. Connection of the Wisp programmer to the 16f877A and 18f4455.

The specialty of the 18f4455 is the USB connection. The result of that is that pin c3 does not exist. Pin 18 is in use for V_{USB}, in the USB projects we will discuss what this is for. Besides pins c4 and c5 can only be used as <u>input</u> and not as output.

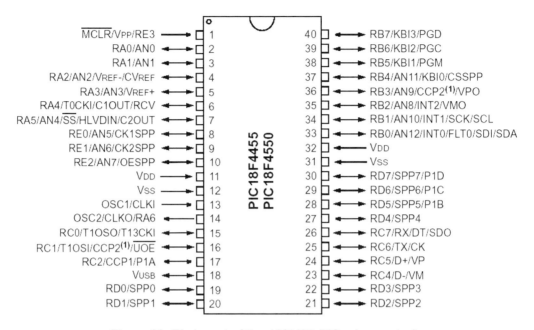

Figure 60. Pin layout of the 18f4455 PIC microcontroller.

We will use a PWM signal. The PWM connections are called CCP1 and CCP2, and you will find them on pins 17 and 16. We will use pin 17 and connect the RC network to it. To see what happens we will connect a LED to the exit of the RC network. Normally we would use a 330 ohm limiting resister, but the RC network eliminates quite a bit of power, in our case this causes an almost 3 volts reduction. For that reason we will use a 100 ohm resistor which results in a current of I = V/R = (2 - 0.7) / 100 = 13 mA. Unfortunately we do not meet the original design criteria which said that the network would be unloaded. Fortunately the current is low so the filter will still work.

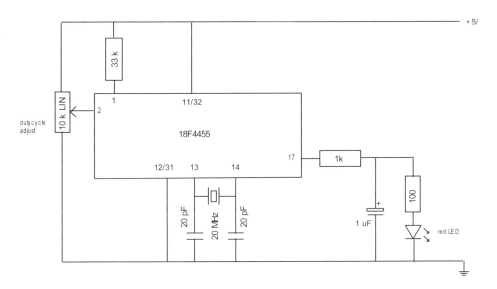

Figure 61. The schematic with a test LED.

The potmeter on pin 2 is used to control the duty cycle of the PWM module.

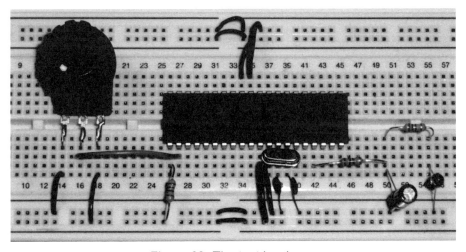

Figure 62. The test hardware.

Software

The default PWM setting in the _bert library is 19,531 kHz. That is excellent for our purpose, because it is way above the cut-off frequency we calculated earlier. This library contains a series of PWM related commands. A full list is shown in the appendix, the next Table shows a small selection.

command	description
PWM_init_frequency (boolean,boolean)	This command can be used to enable the PWM unit or units.
PWM_Set_DutyCycle (number,number)	This command controls the duty cycle. In fact this is the variable that would govern the speed of a motor if it was controlled by PWM.

Table 17. PWM commands used in this project.

The RC network is connected to the first PWM module that can be enabled like this:

    pwm_init_frequency(true,false)

The duty cycle is set with a potmeter on pin 2. The A/D value can be relayed directly to the PWM module because both use bytes in the range 0 to 255.

    resist = adc_read_low_res(0)
    pwm_set_dutycycle(resist,0)

Putting it all together gives this program::

    -- JAL 2.4j
    include 18f4455_bert

    var byte resist

    pwm_init_frequency(true,false)

    forever loop
      -- get variable resistor position
      resist = adc_read_low_res(0)

      -- and use it to control the duty cycle
      pwm_set_dutycycle(resist,0)

    end loop

The exit of the RC network can be connected to the software oscilloscope (WinOscillo) as discussed in section 2.1 so we can observe the filter in action. Do remember to use the special hardware required to connect the project to the PC.

As a first experiment you can vary the position of the potmeter from low to high. You will see that the graph looks the worst about halfway. That is logical because at that point the pulses are exactly as long as the pauses, and that is the most difficult situation to smooth. Never the less the line on the oscilloscope stays reasonably smooth, proof that the filter is working correctly.

The question is of course: did we select the right capacitor. We can easily check this by using different capacitors and observing the effect on the PC screen. In the next Figure you see the results of this test with - from left to right - no filter at all, 100 nF, 220 nF, 470 nF, 1 uF (the value we selected), 4.7 uF and 10 uF.

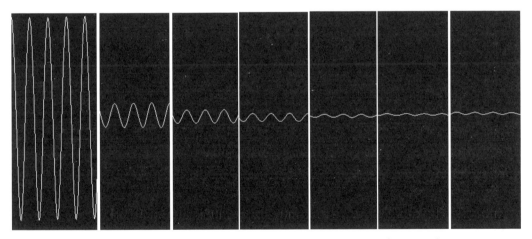

Figure 63. The effect of different capacitors in the RC network.

The graph was made at a duty cycle of 50%, so worst case. Above 1 uF there is not much difference in the smoothness of the graphs. This shows we did select the correct value.

The higher the duty cycle the higher the voltage at the exit of the network. That means that with this technique we have converted a digital value into an analog signal, by sending the digital value as duty cycle to the PWM module. And of course by using an RC network with a 1 k resistor and a 1 uF capacitor to smooth the analog signal.

6.3 Sinus from a Lookup Table

Let's use the technique from the previous chapter to make a beautiful sinus-shaped signal. A sinus shape sounds much more pleasing to the human ear than a square wave that we can make by flipping the pin on a microcontroller from high to low and vise versa.

Technical background

So far we have used separate variables in our projects. It is however also possible to give a whole series of variable the same name. This is called an array.

      var byte Demo[5] = {12,56,100,6,33}

The array shown above is called Demo. So all number in the row are all called Demo. To differentiate between them they are addressed by their position in the row. So Demo[2] for example has value 100. You see that counting - or rather indexing - starts at zero. The number between the square brackets is called index. So demo with index zero is 12. The microcontroller will store this array in RAM memory, and the amount available is limited. If you know that the values in this array will never change you can declare them as constants instead of variables.

      const byte Demo[5] = {12,56,100,6,33}

A constant array is stored in program memory (as would any normal constant). That doesn't make any difference to the way you use the array, except of course you cannot change any of the data. In principle there is no limit to the size of a constant array. Depending on the size of the program an array in an 18f4455 can be up to 12288 bytes long (the size of Program memory). In JAL an array like this is called long table for historical reasons but a more common name is Lookup Table, abbreviated to LUT.

In the download packet you will find an Excel spreadsheet to calculate the sinus for a circle of 360 degrees, or 6.2 rad as Excel likes to call this. A sinus has values in the range of -1 to +1 and that is useless for a microcontroller. So these values must be normalized to a range of 0 to 255. We can do this by multiplying the values (current range -1 to +1) with 127 (range now -127 to +127), and then add 127. The range then becomes 0 to 254.

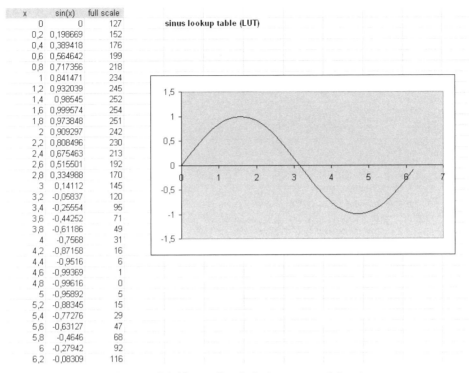

x	sin(x)	full scale
0	0	127
0,2	0,198669	152
0,4	0,389418	176
0,6	0,564642	199
0,8	0,717356	218
1	0,841471	234
1,2	0,932039	245
1,4	0,98545	252
1,6	0,999574	254
1,8	0,973848	251
2	0,909297	242
2,2	0,808496	230
2,4	0,675463	213
2,6	0,515501	192
2,8	0,334988	170
3	0,14112	145
3,2	-0,05837	120
3,4	-0,25554	95
3,6	-0,44252	71
3,8	-0,61186	49
4	-0,7568	31
4,2	-0,87158	16
4,4	-0,9516	6
4,6	-0,99369	1
4,8	-0,99616	0
5	-0,95892	5
5,2	-0,88345	15
5,4	-0,77276	29
5,6	-0,63127	47
5,8	-0,4646	68
6	-0,27942	92
6,2	-0,08309	116

Figure 64. Normalized sin in a spreadsheet.

The values we have made this way must now be stored in the JAL program as long table, so a constant array.

> const byte sinus[32] = {127,152,176,198,218,233,245,252,253,250,242,229,212, 192,169,144,119,94,70,49,30,16,6,0,0,5,14,28,46,67,91,116}

The next step is to collect these value one by one and send them to the PWM unit as duty cycle at a very steady pace. Just like in section 6.1 we will use the timer0 interrupt. The tables in that section are for the 16f877A, and in this project we use the 18f4455. This microcontroller is much faster, and uses slightly different register names.

REGISTER 11-1: T0CON: TIMER0 CONTROL REGISTER

R/W-1	R/W-1	R/W-1	R/W-1	R/W-1	R/W-1	R/W-1	R/W-1
TMR0ON	T08BIT	T0CS	T0SE	PSA	T0PS2	T0PS1	T0PS0
bit 7							bit 0

Legend:			
R = Readable bit	W = Writable bit	U = Unimplemented bit, read as '0'	
-n = Value at POR	'1' = Bit is set	'0' = Bit is cleared	x = Bit is unknown

bit 7   TMR0ON: Timer0 On/Off Control bit
         1 = Enables Timer0
         0 = Stops Timer0

bit 6   T08BIT: Timer0 8-Bit/16-Bit Control bit
         1 = Timer0 is configured as an 8-bit timer/counter
         0 = Timer0 is configured as a 16-bit timer/counter

bit 5   T0CS: Timer0 Clock Source Select bit
         1 = Transition on T0CKI pin
         0 = Internal instruction cycle clock (CLKO)

bit 4   T0SE: Timer0 Source Edge Select bit
         1 = Increment on high-to-low transition on T0CKI pin
         0 = Increment on low-to-high transition on T0CKI pin

bit 3   PSA: Timer0 Prescaler Assignment bit
         1 = TImer0 prescaler is NOT assigned. Timer0 clock input bypasses prescaler.
         0 = Timer0 prescaler is assigned. Timer0 clock input comes from prescaler output.

bit 2-0   T0PS2:T0PS0: Timer0 Prescaler Select bits
         111 = 1:256 Prescale value
         110 = 1:128 Prescale value
         101 = 1:64   Prescale value
         100 = 1:32   Prescale value
         011 = 1:16   Prescale value
         010 = 1:8   Prescale value
         001 = 1:4   Prescale value
         000 = 1:2   Prescale value

Figure 65. The t0con register of the 18f4455 microcontroller.

Setting the interrupt frequency of the 18f4455 is done in the t0con register (in the 16f877A this was the option_reg register). A special property of this particular microcontroller is that it increases the frequency of the crystal. Just like the 16f877A the 18f4455 is connected to a 20 MHz crystal, but the 18f4455 transforms this internally to 48 MHz. Let's do some math. The microcontroller uses an internal clock frequency of 48 MHz. Every <u>fourth</u> pulse timer0 is incremented by one (because each instruction requires four clock pulses). An interrupt is generated when timer0 overflows, so after 256 additions - or increments. That means the default interrupt frequency is:

$$48{,}000{,}000 / 4 / 256 = 46{,}875 \text{ Hz}$$

The 18f4455 timer0 can also be used as a 16 bit timer, That means the interrupt frequency a significantly lower because timer0 will overflow after 65,536 increments:

48.000.000 / 4 / 65,536 = 183 Hz

This results in the following interrupt Table.

Prescaler	TMR0 8 bit Frequency (Hz)	TMR0 16 bit Frequency (Hz)
not assigned	46,875	183
0	23,438	92
1	11,719	46
2	5,859	23
3	2,930	11
4	1,465	5.7
5	732	2.9
6	366	1.4
7	183	0.7

Table 18. Interrupt frequencies for the 18f4455 microcontroller.

Let's assume we use an interrupt frequency of 11,719 Hz. There are 32 values in the LUT that together form a single sound wave. That means that the sound frequency would be 11,719/32 = 366 Hz. The audible frequency range is normally speaking between 20 Hz and 20 kHz so this will be a good interrupt frequency to use. So in t0con we need to enable the timer, use 8 bits, and a prescaler of 1.

t0con = 0b_1100_0001

The intcon register is (by coincidence!) identical to the 16f877A, so we can copy that setting.

intcon = 0b_1010_0000

INTCON REGISTER (ADDRESS 0Bh, 8Bh, 10Bh, 18Bh)

R/W-0	R/W-0	R/W-0	R/W-0	R/W-0	R/W-0	R/W-0	R/W-x
GIE	PEIE	T0IE	INTE	RBIE	T0IF	INTF	RBIF

bit 7   bit 0

- **bit 7**   **GIE:** Global Interrupt Enable bit
  - 1 = Enables all unmasked interrupts
  - 0 = Disables all interrupts
- **bit 6**   **PEIE:** Peripheral Interrupt Enable bit
  - 1 = Enables all unmasked peripheral interrupts
  - 0 = Disables all peripheral interrupts
- **bit 5**   **T0IE**: TMR0 Overflow Interrupt Enable bit
  - 1 = Enables the TMR0 interrupt
  - 0 = Disables the TMR0 interrupt
- **bit 4**   **INTE**: RB0/INT External Interrupt Enable bit
  - 1 = Enables the RB0/INT external interrupt
  - 0 = Disables the RB0/INT external interrupt
- **bit 3**   **RBIE**: RB Port Change Interrupt Enable bit
  - 1 = Enables the RB port change interrupt
  - 0 = Disables the RB port change interrupt
- **bit 2**   **T0IF**: TMR0 Overflow Interrupt Flag bit
  - 1 = TMR0 register has overflowed (must be cleared in software)
  - 0 = TMR0 register did not overflow
- **bit 1**   **INTF**: RB0/INT External Interrupt Flag bit
  - 1 = The RB0/INT external interrupt occurred (must be cleared in software)
  - 0 = The RB0/INT external interrupt did not occur
- **bit 0**   **RBIF**: RB Port Change Interrupt Flag bit
  - 1 = At least one of the RB7:RB4 pins changed state; a mismatch condition will continue to set the bit. Reading PORTB will end the mismatch condition and allow the bit to be cleared (must be cleared in software).
  - 0 = None of the RB7:RB4 pins have changed state

Figure 66. The intcon register of the 18f4455 microcontroller.

Hardware

The hardware is in fact identical to the previous project, but without the LED. After the RC filter the project is connected to the PC - using the special hardware from section 2.2 - so we can see the result in WinOscillo, and hear it through the PC speakers. WinOscillo is after all connected to the microphone connection on our PC.

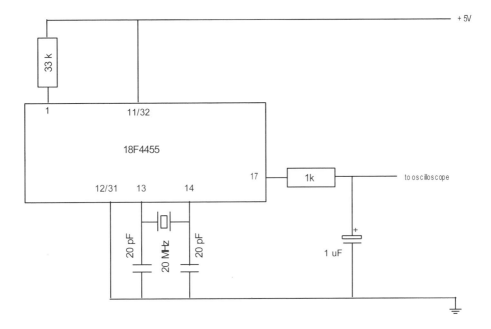

Figure 67. LUT schematic with RC filter.

On the right side you can see the wire leading to the oscilloscope hardware and from there to the PC.

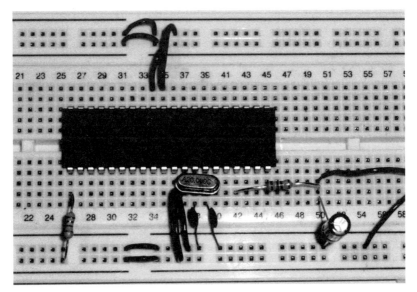

Figure 68. The sinus project on a breadboard.

105

Software

In the interrupt routine the program must fetch a number from the LUT and pass it to the PWM module as duty cycle. In general you should keep an interrupt routine short, to prevent the next interrupt from arriving while the previous one is still being processed. After the 32nd number from the LUT - the counter is then 31 because we start with 0 - the counter should restart. This would be a good way to do that:

```
counter = counter +1
if counter == 32 then
   counter = 0
end if
```

Optionally you could use modulo. Modulo is the remainder of a division. So 9 modulo 2 equals 1 - because 9/4 equals 2 with remainder 1. Modulo is written as percentage sign:

```
counter = (counter+1)%32
```

This single line does the same as the previous group of lines, so that will speed things up a bit. It increments the counter with 1, and restarts after 31. If you find this difficult to understand take a piece of paper and do the math manually. Start with the counter at 29 (lower numbers aren't very interesting) and see what happens. If things get complicated I try it out this way too.

The sin that belongs to index counter is fetched from the LUT.

```
sinus[counter]
```

We might as well enter it directly into the PWM duty cycle command pwm_set_dutycycle. The next step is to reset the timer0 interrupt flag t0if. And then we wait for the next interrupt. The completed interrupt procedure (called next_number) looks like this:

```
procedure next_number is
  pragma interrupt

  if t0if then
    counter = (counter+1)%32
    pwm_set_dutycycle(sinus[counter],0)
    t0if = 0
  end if
end procedure
```

The remainder of the program consists of declarations, the LUT itself and a main program that doesn't do anything.

```
-- JAL 2.4j
include 18f4455_bert

-- variables and lut
const byte sinus[32] = {127,152,176,198,218,233,245,252,253,250,242,229,212,
192,169,144,119,94,70,49,30,16,6,0,0,5,14,28,46,67,91,116}
var byte counter

-- enable pwm
pwm_init_frequency(true,false)

-- timer0 interrupt routine
procedure next_number is
  pragma interrupt

  if t0if then
      -- if timer0 interrupt then read next value from the
      -- lut table and set the duty cycle to this value, use
      -- mod32 to keep the counter in range
      counter = (counter+1)%32
      pwm_set_dutycycle(sinus[counter],0)
      t0if = 0
  end if
end procedure

-- enable timer0, set as 8 bits, use prescaler 1
t0con = 0b_1100_0001

-- enable interrupts and timer0 interrupt
intcon = 0b_1010_0000

-- main loop, do nothing but keep the program running
forever loop
end loop
```

WinOscillo shows a very satisfying result. The curve has a good sinus shape with a very small serration at the peaks. And the sound coming from the PC speakers sounds good.

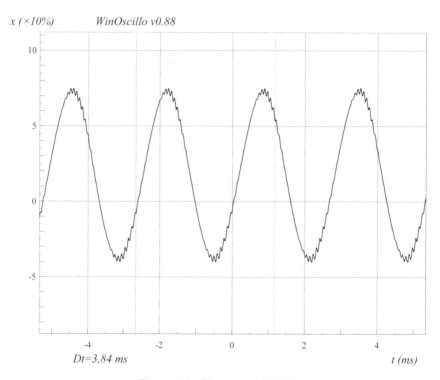

Figure 69. Sinus met RC filter.

If you are like me you are now wondering what the effect of the RC filter is on this graph. You can easily see that by connecting WinOscillo before instead of after the filter. You will then get the result shown in the next Figure. A staggering difference, the RC network is performing great!

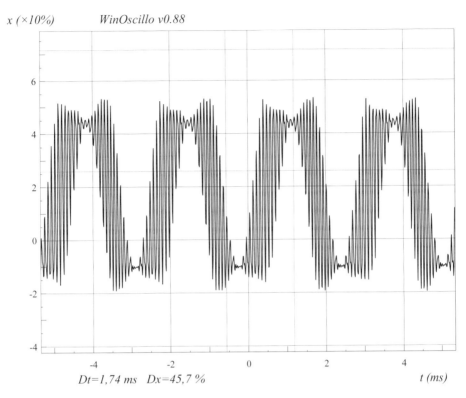

Figure 70. Sinus without the RC filter.

6.4 Siren with amplifier

When a loudspeaker is connected directly to the microcontroller the volume is rather low. In this project we will use an amplifier to make a loud siren.

Technical background

We will use the integrated TDA1011 amplifier manufactured by Philips. The hardware setup is copied from the datasheet with some alterations. Besides I have added a 1 k linear potmeter to control the volume.

The amplifier can deliver 1 to 6.5 W of power depending on the voltage of the power supply. The minimum voltage is 3.6 volts, the maximum 20. According to the datasheet this is the relationship between power and voltage

voltage (volt)	power (watt)
6	1
9	2.3
12	4.2
16	6.5

Table 19. Relationship between supply voltage and delivered power.

It is not clear how much power the unit delivers at 5 volts.

Hardware

In this project a 16f877A microcontroller is used, obviously at 5 volts. If you want more power you can run the microcontroller and the amplifier on different voltages. The ground of both power supplies must be connected to each other. The amplifier, the 220 nF capacitors and the 10 uF capacitor are connected to the higher voltage. In that case it is probably better to screw the amplifier IC to a good size heatsink.

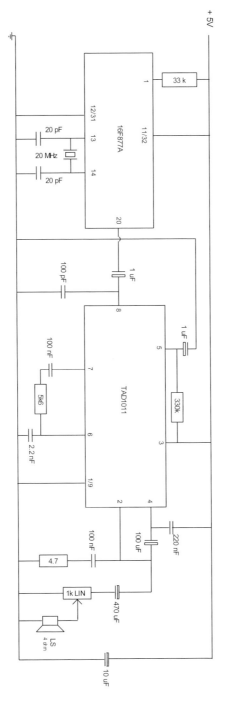

Figure 71. Microcontroller with 1 W amplifier.

This is the project on a breadboard connected to a 4 ohm 10 W Dual loudspeaker. The volume is rather loud, and the voltage regulator on the Wisp648 is running a bit warm. If you want to use this setup over a longer period of time a heatsink on the 7805 of the Wisp648 is definitely required.

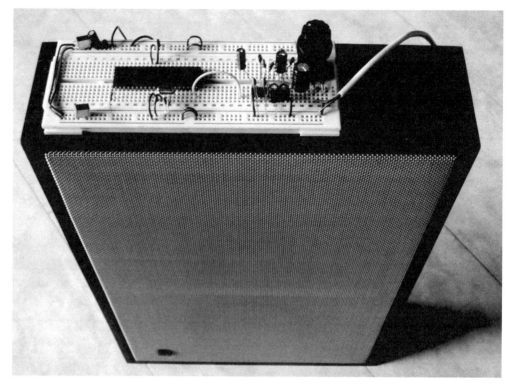

Figure 72. The siren connected to a 4 ohm / 10 W Dual loudspeaker.

Software

Dutch police sirens used to be dual tone with sound frequencies 375 and 500 Hz. The idea behind that was that the high tone would be audible over a long distance outside, while the low tone would be able to penetrate a vehicle.

So far when we wanted to make sound we used an interrupt. This time we will use a different technique. That makes the program somewhat simpler, but the disadvantage is that the microcontroller doesn't have time to do anything else than make a sound.

A tone of 375 Hz means a sound wave of $1/375 = 2.7 \; 10^{-3}$ seconds and an vibration frequency of $2.7 \; 10^{-3} / 2 = 1.3 \; 10^{-3}$ seconds. A vibration is the flipping of a pin. The closest higher standard delay command is 2 mS, and since we are not very concerned about accuracy we will use it. As discussed before we will use an intermediate variable called flag to change the position of the pin. This time a large capacitor is connected to the pin - at the input of the amplifier - so chances are that the pin will not be able to follow the rapid flipping.

```
for 250 loop
   delay_1ms(2)
   flag = !flag
   pin_d1 = flag
end loop
```

Following that same logic we can calculate that for 500 Hz the delay per vibration is 1 mS. The delay is half as long as for the low frequency sound. That means that the 500 Hz tone will sound half as long. We correct that by doubling the number of times that the program runs that particular loop. - 500 times versus 250 times for the 375 Hz tone).

```
-- JAL 2.4i
include 16F877A_bert

--define variable
var bit flag

-- output pin
pin_d1_direction = output

forever loop

   -- high tone
   for 500 loop
     delay_1ms(1)
     flag = !flag
     pin_d1 = flag
   end loop

   -- low tone
   for 250 loop
     delay_1ms(2)
     flag = !flag
     pin_d1 = flag
   end loop

end loop
```

6.5 A talking 18F4685

In this project we will make a microcontroller speak a sentence.

Technical background

In principle this program is very simple: a short sound is stored in the microcontroller, and played back using the PWM module. Sound can be stored in many different ways on a PC. Best known at the moment is MP3, because this technique allows for a serious compression thus reducing the required amount of space. This compression also means that the sound quality is reduced. Considering the way that MP3's are generally played back - using small earplugs as loudspeakers - this is not regarded as a big issue.

In this project we will use a different way of storing sound file, the so called WAV format. A WAV file is made by using an analog to digital converter (just like the one in the microcontroller) to turn the music into digital numbers. These numbers can be send to the PWM unit to be transformed back into an analog signal using the technique described in section 6.2.

WAV files are not compressed and thus rather large. The only place where we could possibly store large files is in program memory as a Lookup Table. The faster we send the WAV data to the PWM unit the better the sound quality, but also the more data we need per second, and thus the larger the amount of data. For that reason we use the 18f4685 microcontroller. This is a 40 pin microcontroller with a large program memory, as you can see in the next table where we compare the three microcontrollers that we have used so far in this book.

PIC	Program (words)	RAM (bytes)	EEPROM (bytes)	I/O pins	Analog inputs	RS232	USB	I^{2}C	SPI	PWM	CAN bus	Speed (mips)
16f877A	8192	368	256	33	8	2	0	2	2	2	0	5
18f4455	12288	2048	256	35	13	2	1	2	2	2	0	12
18f4685	49152	3328	1024	36	11	2	0	2	2	2	1	10

Table 20. Comparison between three microcontrollers.

The program memory of this microcontroller is 49,152 words. The numbers in the WAV file are bytes, and we can fit two bytes in one word. That means we can store 98,304 bytes, minus the amount of room that the program itself needs. We will use an interrupt frequency of 9,766 Hz. The sound quality is acceptable, and we can fit 98,304 / 9,766 = 10 seconds of speech in memory.

Setting the interrupt is done using the intcon and t0con registers just like in the 18f4455. The 18f4685 does not have a mechanism to internally increase the clock speed so it runs at 20 MHz, just like the crystal. We do have the option of setting timer0 to 8 or 16 bits, which results in the following interrupt frequency table.

Prescaler	TMR0 8 bit Frequency (Hz)	TMR0 16 bit Frequency (Hz)
not assigned	19.531	76
0	9.766	38
1	4.883	19
2	2.441	9,5
3	1.221	4,8
4	610	2,3
5	305	1,2
6	153	0,6
7	183	0,3

Table 21. Interrupt frequencies for the 18f4685 microcontroller.

For an interrupt frequency of 9,766 Hz the registers must be set as follows:

    t0con = 0b_1100_0000
    intcon = 0b_1010_0000

This means that the sound is played with a speed of 9,766 bits per second, also known as the bitrate. Since the sound needs to be preprocessed on the PC we need to record it in a WAV file with a bitrate of 9,766 Hz. That is less simple than it appears. Many programs are capable of recording with that bitrate - also called sampling rate - but when the file is saved they still use a more common bitrate. Cool Edit can do this, but this is hardly a freeware program[27]. In the download package you will find a file that is recorded and saved at the correct bitrate and converted to the proper JAL format: hello9766hzjmp.jal. It contains the text "Hello, I'm your microcontroller". At the end of this project you will find instructions as to how you can record your own sound and use it in the microcontroller.

---

[27] On the internet you will still be able to find old shareware versions.

Hardware

The amplifier that we use in this project is identical to the one in the previous project. This time however it is connected to pin c2 - the PWM pin of unit 1. Another change is that we use the 18f4685 microcontroller that has a lot of memory, and the capability for CANbus. We will discuss CANbus in more detail in chapter 9.

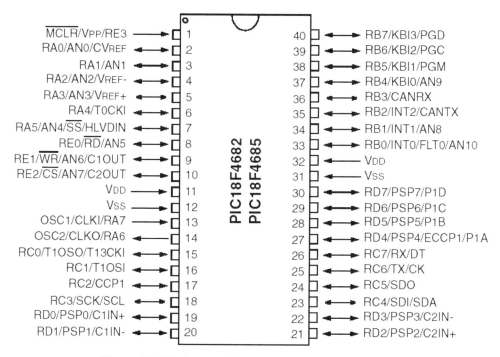

Figure 73. Pin layout of the 18f4685 microcontroller.

The Wisp programmer is connected to the same pins as for the 18f4455.

PIC	yellow	blue	green	white	red	black	jumper
16f877A	1	40	39	36	+5V	0	no
18f4455	1	40	39	38	+5V	0	no
18f4685	1	40	39	38	+5V	0	no

Table 22. Connection of the Wisp programmer.

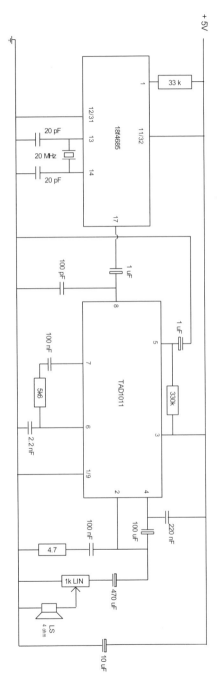

Figure 74. Schematic of the talking microcontroller.

This time the 4 ohm 10 W Dual loudspeaker is not shown in the picture, just the wires on the right hand side leading to it.

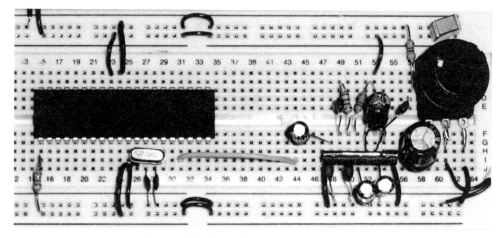

Figure 75. The talking microcontroller.

Software

The sound is in a LUT in a different file called hello9766hzjmp.jal. You don't need to copy the data from this file into the main program. Instead we use the "include" command

    -- JAL 2.4j
    include 18f4685_bert
    include hello9766hzjmp

This command tells the JAL compiler that at this point everything that is in that file must be inserted. The use of includes makes programs easier to read and maintain. An include file <u>must</u> have the extension .JAL, and you <u>must not</u> use that extension in the include statement. Apart from that the include must be in the same directory as the JAL program itself, or in the directory where you keep all your libraries. If not you will get an error message.

In the program in section 6.3 we used an interrupt to fetch a number from the LUT and send it to the PWM module. This time we will use a slightly different technique which keeps the interrupt routine small. Professional programmers therefor prefer this technique to the previous one. The main program will fetch the next number from the LUT and than wait for the interrupt to happen in a while loop.

```
-- retrieve values from long table
data = long_table[counter]

-- wait for the interrupt
while ! interrupted loop end loop
interrupted = false
```

The variable "interrupted" is used as a flag. The main program will make this variable false, and then wait for it to become true. This can only happen in the interrupt routine. Immediately after the variable has become true the data is send to the PWM module, therefor the speed is exactly identical to the interrupt speed. Using prefetching allows us to correct for the difference it may take the microcontroller between fetching data from the beginning or the end of memory.

```
-- JAL 2.4j
include 18f4685_bert
include hello9766hzjmp

--define variable
var bit interrupted
var byte data
var word counter

-- output pin
pin_d1_direction = output

-- the interrupt routine
procedure playback is
  pragma interrupt

  if t0if then
    -- signal that the interrupt has occurred
    interrupted = true

    -- clear t0if to re-enable timer0 interrupts
    t0if = 0
  end if
end procedure

-- enable pwm module
pwm_init_frequency(true,false)
```

```
-- enable timer0 frequency 9766 Hz
t0con = 0b_1100_0000

-- enable interrupts
intcon = 0b_1010_0000

forever loop

    -- start at the beginning of the array
    counter = 0

    for count(long_table) loop

        -- retrieve values from long table
        data = long_table[counter]

        -- wait for the interrupt
        while ! interrupted loop end loop
        interrupted = false

        -- send data to pwm module
        pwm_set_dutycycle(data,0)

        -- increase counter for the next value from the long table
        counter = counter + 1

    end loop

    -- if the end of data is reached wait a bit
    delay_1s(2)

end loop
```

When you compile this program you will get the following notification:

```
jal 2.4j (compiled Mar 12 2009)
0 errors, 0 warnings
Code area: 9162 of 98304 used
Data area: 20 of 3328 used
Software stack available: 2816 bytes
Hardware stack depth 3 of 31
```

If the LUT is only 1 byte long the size of the program is 198 (try this to see if I'm correct). In reality the LUT contains 17,916 bytes. These are stored two by two so that requires 8,958 words. The expected required space program memory - the code area - is therefor 198+8,958 = 9,156. The true value is 9,162, so 6 words larger. This is caused by the overhead in the program. As you can see there is plenty of room for a much longer sound.

Use your own sound

If you want to use your own sound in the microcontroller you need to follow these steps:

1. Record a suitable sound of no more than 10 seconds. The conversion to 9,766 Hz causes a serious loss of volume and high pitch, so make sure the sound is recorded as loud as possible and with plenty of high pitch. Save it in a directory and with a file name that contain no spaces.

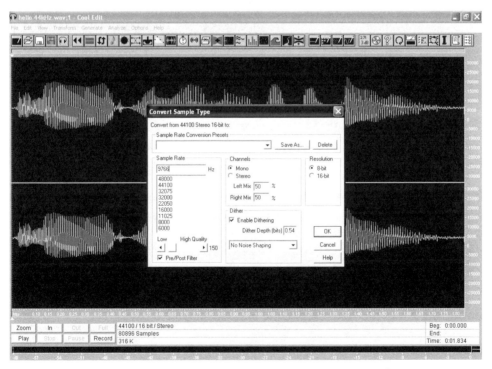

Figure 76. Conversion of the sampling rate to the interrupt frequency.

2. Use Cool Edit to convert the sample rate to 9,766 Hz. In the menu select "edit" - "convert sample rate". As settings use: no presets, 9,766 Hz, mono, 8 bits. Save the file

using a different name. Tip: add the text 9766Hz to the filename, and again make sure to use no spaces.

3. Use the program WAVconvert in the download package to convert the file to a long table. Enter the filename in the lower left corner and then click on the "View" button. Verify that the program shows the correct bitrate. Then select "raw" - "long table" - "normal" and click on the "Go!" button. The program will now produces a series of files in the directory of the original sound file. The file that we need for this project has JMP as extension. Copy that file to the directory where your JAL program is, and change the extension to JAL.

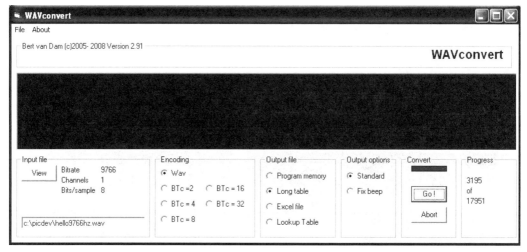

Figure 77. Conversion of the sound file to a long table.

4. Open the file in JALedit and check if the file contains one or more zeros at the very end. If that is the case they must be removed, and you need to modify the length of the table to account for the removed data. If for example you deleted four zero's than you need to reduce the xxxx in the first line of the program by four as well.

    const byte long_table[xxxxx] = {

If you do not remove the zeros they sound like an annoying tick.

5. In the main program add an include with the name of this file.

6. Compile the program and check to see if it fits in memory. If yes then you can transfer the file to the 18f4685 microcontroller. Playback of your soundbit will start automatically and will be repeated every two seconds.

7 Process sound

In this chapter we will discuss a few projects that deal with processing sound. We will use sounds generated on a PC and fed directly into our project, but we will also use a microphone, both with and without amplifier.

7.1 Comparator

In virtually all sound processing projects we use a comparator. So before we start with these projects we will first do a small project to clarify what the comparator does and how it works. This project by the way is not sound related.

Technical background

The 16f877A has two comparators. These are built-in components that can compare two voltages. This is not done in software but in hardware so it is lightning fast. A comparator is particularly useful for applying a threshold value. Only when a voltage is above - or below - that threshold the microcontroller will react. If you are into traditional electronics: this is a lot like a Schmitt trigger.

In this project we will use a threshold value of 2.5 volts by connecting two 10 k resistors in series between the ground and the +5 volts, and connect one pin of the comparator to the middle. The total current going through these resistor is $I = V/R = 5/20k = 0.25$ mA. That means the voltage over one resistor is $V = I * R = 0.25$ mA $* 10k = 2.5$ volts. The other pin of the comparator will be connected to a potmeter, which can supply any voltage between 0 and +5 volts. Depending on the position of the knob the comparator will be "on" or "off".

The cmcon register, bits 2 to 0, are used to set the comparators. Because bit 7 is always on the left side we call these bits 2 to 0 and not bit 0 to 2, which might seem more logical at first glance. We generally write this down as bit 2:0.

CMCON REGISTER

R-0	R-0	R/W-0	R/W-0	R/W-0	R/W-1	R/W-1	R/W-1
C2OUT	C1OUT	C2INV	C1INV	CIS	CM2	CM1	CM0
bit 7							bit 0

bit 7   **C2OUT**: Comparator 2 Output bit
When C2INV = 0:
1 = C2 V$_{IN+}$ > C2 V$_{IN-}$
0 = C2 V$_{IN+}$ < C2 V$_{IN-}$
When C2INV = 1:
1 = C2 V$_{IN+}$ < C2 V$_{IN-}$
0 = C2 V$_{IN+}$ > C2 V$_{IN-}$

bit 6   **C1OUT**: Comparator 1 Output bit
When C1INV = 0:
1 = C1 V$_{IN+}$ > C1 V$_{IN-}$
0 = C1 V$_{IN+}$ < C1 V$_{IN-}$
When C1INV = 1:
1 = C1 V$_{IN+}$ < C1 V$_{IN-}$
0 = C1 V$_{IN+}$ > C1 V$_{IN-}$

bit 5   **C2INV**: Comparator 2 Output Inversion bit
1 = C2 output inverted
0 = C2 output not inverted

bit 4   **C1INV**: Comparator 1 Output Inversion bit
1 = C1 output inverted
0 = C1 output not inverted

bit 3   **CIS**: Comparator Input Switch bit
When CM2:CM0 = 110:
1 = C1 V$_{IN-}$ connects to RA3/AN3
    C2 V$_{IN-}$ connects to RA2/AN2
0 = C1 V$_{IN-}$ connects to RA0/AN0
    C2 V$_{IN-}$ connects to RA1/AN1

bit 2   **CM2:CM0**: Comparator Mode bits
Figure 12-1 shows the Comparator modes and CM2:CM0 bit settings.

Figure 78. The cmcon register of the 16f877A microcontroller.

There are numerous different ways to connect and enable the comparators in the 16f877A microcontroller, and you will find these in section 21.1 of the datasheet of the 16f877A. In this project we select setting 001.

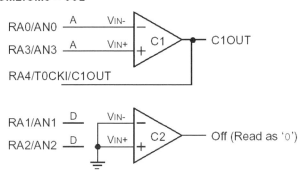

Figure 79. Comparator setting 001.

This setting means that comparator C2 is disabled. Pins a1 and a2 function normally. Comparator C1 is enabled. That means that the voltage on pins a0 and a3 will be compared to each other, and the result is presented in variable c1out and on pin a4.

What that result is depends on the settings in cmcon. If we select zero for c1inv (bit 4 of cmcon) then c1out (bit 6 of cmcon) will get the following values:

input	c1out
$V_{IN+} > V_{IN-}$ (so the voltage on pin a3 > a0)	1
$V_{IN+} < V_{IN-}$ (so the voltage on pin a3 < a0)	0

Table 23. Effect of V_{IN+} and V_{IN-} on c1out.

Bit 6 of cmcon is known to JAL as c1out because the 16f877A_bert library is loaded, so you can use this variable in the JAL program.

Hardware

Pin 2 (a0) is set on 2.5 volts by a voltage divider made out of two 10 k resistors in series. Pin 5 (a3) has a variable voltage due to the potmeter. The way the LED is connect to pin a4 may surprise you, because a second resistor is added from the pin to the +5 volts. Pin a4 is a very special pin. In chapter 5 we have seen that this pin can be connected to (relatively) high AC voltages. Another specialty of this pin is that it is incapable of making itself high. The pin can only be made low (0 volts) but not high (+5 volts). The 1k resistor is called a pull-up resistor because it pulls the pin high when it is not actively making itself low. During that time the LED gets its current through the 1 k resistor and the 330 ohm resistor. If the microcontroller wants to switch the LED off it makes the pin low. At that moment both sides of the LED are low so it is off. There is at that moment of course a current through the 1 k resistor: I = V/R = 5/1000 = 5 mA. The microcontroller can handle that easily.

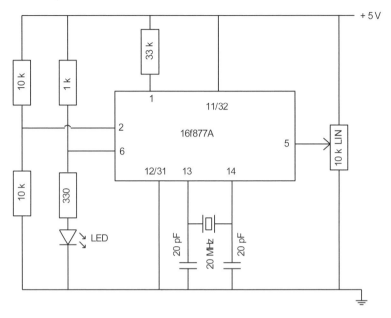

Figure 80. Comparator test schematic.

The next Figure shows the project on the breadboard.

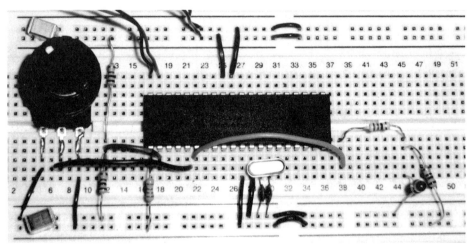

Figure 81. The Wisp still connected, without the yellow and white wires.

The Wisp needs to stay connected after programming (if you want you may disconnect the white and yellow wires) because we will use the passthrough functionality of the programmer. The programmer is capable of making a direct RS232 connection between the microcontroller and the PC. That means you can communicate with the microcontroller. This is a unique property of this programmer.

Software

The Wisp makes a serial connection, so on the PC you need to run a terminal or communications program. The program that we use in this project, and very often in other projects in this book, is MICterm. This program is especially designed for use with microcontrollers, and is part of the download package.

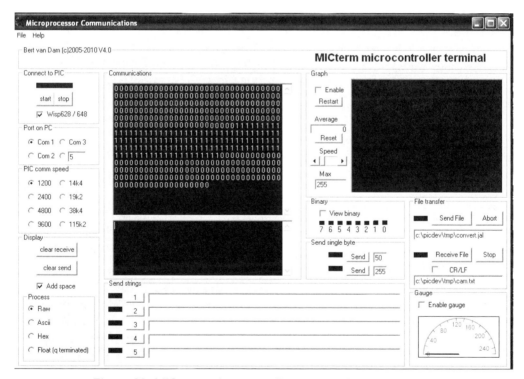

Figure 82. MICterm microcontroller communication program.

In order to switch the Wisp in passthrough mode a special command has to be given. You can issue this command by checking the Wisp62/6489 checkbox in MICterm, prior to clicking on the Start button. Make sure to select the correct COM port - the one where the Wisp is connected - and select a speed of 1200 baud.

In the microcontroller a special software RS232 library is loaded by default (as part of the _bert library) which allows you to use the following commands in your JAL program. This does not apply to the 10f200 microcontroller, which is too small for this functionality.

Command	Description
Serial_SW_Read(data)	Receive data and put it in the variable data.
Serial_SW_Write (data)	Send the contents of the variable data.
Serial_SW_Locate(horizontal, vertical)	Move the cursor (on a VT52 terminal or emulation) to the coordinates horizontal, vertical.
Serial_SW_Clear	Clear the screen from the current cursor position (on a VT52 terminal or emulation).
Serial_SW_Home	Move the cursor to the home position - upper left corner (on a VT52 terminal or emulation).
Serial_SW_Byte(data)	Send a byte named data as three digits - or less- instead of a single number.
Serial_SW_Printf(array)	Send a complete array - which needs to be defined first - with a single command. For example const byte mystr[] = "Bert van Dam" followed by serial_sw_printf(mystr).

Table 24. A selection of serial software commands.

In this program we only use one command: serial_sw_write. Using this command you can check quite easily if a program is doing what you want it to do. Put this command in your program and use it to send the content of a variable that you are interested in to the PC. Now start MICterm, and observe how this variable changes. If the project is not connected to a PC the program will still run; the serial communications command will not cause it to "hang". An ideal way of debugging a program that I use very often. One of the reasons for me to recommend this programmer!

The program itself is rather straightforward.

```
-- JAL 2.4i
include 16f877A_bert

-- define the pins
pin_a0_direction = input
pin_a3_direction = input
pin_a4_direction = output

-- set single comparator
cmcon = 0b_0000_0001

forever loop

    -- send the data to the PC, c1out is the output which also goes to RA4
    serial_sw_write(c1out)

end loop
```

Once the program is running the LED should be on if the potmeter is turned to the right, and off when it is turned to the left. Optionally you can use MICterm to observe the value of variable c1out. If the LED is on the value should be 1, otherwise the value is 0. Make sure to select "raw" in MICterm otherwise you will see the ASCII values of 0 and 1 which are unfortunately both unprintable characters.

7.2 Sound switch

In this project we will built a sound switch: a switch that can be controlled by sound. You can for example light a LED by clapping your hands. If a LED is not impressive enough you can combine this project with for example project 4.4 and light an mains light bulb.

Technical background

In the previous program we used two resistors to make a reference voltage of 2.5 volts. In this project we will use the reference voltage to set the sensitivity of the sound switch. A sound that generates a voltage below the reference voltage will not be heard. The reference voltage - now in fact the sound threshold - must be adjustable in order to achieve this. We could use a potmeter for this but that would mean only a tiny fraction of the potmeter range would be used. A better solution is to let the microcontroller set its own reference voltage, and use the potmeter to control that process. That way we can use much more of the range of the potmeter, making the sensitivity easier to adjust. Just like in the previous project we use comparator setting 001.

    cmcon = 0b_0000_0001

The next step is to enable the unit that can take care of an internal reference voltage, the so called Comparator voltage Reference Generator, or CVR. This unit is controlled by the cvrcon control register. JAL is familiar with all bits from this register.

CVRCON CONTROL REGISTER (ADDRESS 9Dh)

R/W-0	R/W-0	R/W-0	U-0	R/W-0	R/W-0	R/W-0	R/W-0
CVREN	CVROE	CVRR	—	CVR3	CVR2	CVR1	CVR0
bit 7							bit 0

bit 7	**CVREN**: Comparator Voltage Reference Enable bit	
	1 = CVREF circuit powered on	
	0 = CVREF circuit powered down	
bit 6	**CVROE**: Comparator VREF Output Enable bit	
	1 = CVREF voltage level is output on RA2/AN2/VREF-/CVREF pin	
	0 = CVREF voltage level is disconnected from RA2/AN2/VREF-/CVREF pin	
bit 5	**CVRR**: Comparator VREF Range Selection bit	
	1 = 0 to 0.75 CVRSRC, with CVRSRC/24 step size	
	0 = 0.25 CVRSRC to 0.75 CVRSRC, with CVRSRC/32 step size	
bit 4	**Unimplemented**: Read as '0'	
bit 3-0	**CVR3:CVR0**: Comparator VREF Value Selection bits 0 ≤ VR3:VR0 ≤ 15	
	When CVRR = 1:	
	CVREF = (VR<3:0>/ 24) • (CVRSRC)	
	When CVRR = 0:	
	CVREF = 1/4 • (CVRSRC) + (VR3:VR0/ 32) • (CVRSRC)	

Figure 83. The cvrcon control register of the 16f877A.

Of course the unit must be switched on (cvren = 1) and the result must be sent to a pin connected to the comparator, so a0 of a3. Unfortunately that option doesn't exist so we will send the output to pin a2 (cvroe = 1) and then later use a wire to connect pin a2 to pin a3. The next question is which range we want to use. The description at bit 5 is a bit complicated, but it basically boils down to the question if we want to control the voltage starting at zero, or not. The microphone delivers a voltage in the 10 mV range, so very low. That means in our case the answer is "yes", and we select 1 for cvrr.

What the voltage will be is governed by bits 3:0. If they are all zero the voltage is 0/24 * 5 = 0 volts. If they are all one, then the voltage is 15/24 * 5 = 3.1 volts. In between the voltage is adjustable in steps of about 0.2 volts.

Hardware

The design of the microphone part of this project is by Patrick Frought[28]. The Electret microphone is connected using a standard network to the power and the input of the comparator. The other comparator input is connected to the output of the CVR unit. The 10 k resistor connects both inputs to each other. When the microphone detects a sound it will deliver voltage swings in the 10 mV range that pass through the capacitor. Because of the 10 k resistor the effect of these swings is larger in pin a0 than in pin a3 thus causing the comparator to be more sensitive.

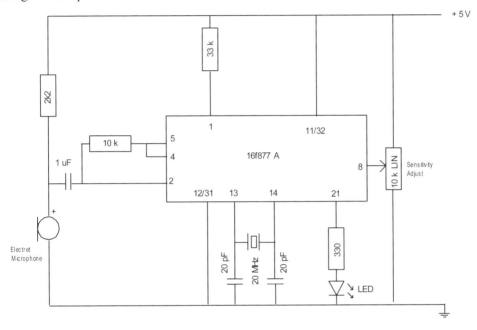

Figure 84. Schematic of the sound switch.

Please note that Electret microphones have a plus and a ground side. If you take a close look at the bottom of the microphone you will notice that one of the pins is connected to the house itself. This is the pin that is connected to the 0 in the schematic.

---

[28] Suggested using the JAL usergroup.

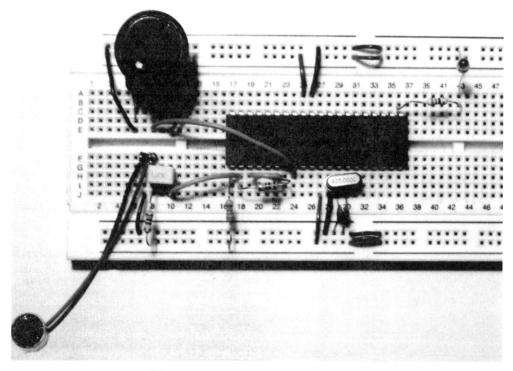

Figure 85. Sound switch in operation.

Software

Most parts of the software have already been discussed in the technical background section. This part however is new:

> sensitivity = ADC_read_low_res(5)
> sensitivity = sensitivity/17
> CVR0 = (sensitivity & 0b_0000_0001)
> CVR1 = (sensitivity & 0b_0000_0010) >> 1
> CVR2 = (sensitivity & 0b_0000_0100) >> 2
> CVR3 = (sensitivity & 0b_0000_1000) >> 3

The first step is to read the A/D converter on channel 5 (pin 8). This results in a value between 0 - 255. In the next step the range is reduced to 0 - 15. The & sign is the symbol for AND. AND means that a bit will remain set if it is set on both sides.

Let's take a look at the truth table for AND.

number 1	number 2	number 1 & number 2
1	1	1 & 1 = 1
1	0	1 & 0 = 0
0	1	0 & 1 = 0
0	0	0 & 0 = 0

Table 25. AND truth table.

If for example the value of sensitivity is 3, then the first AND operation then becomes

3 AND 0b_0000_0001 = ?

Using the Windows calculator you can convert three to binary. Then the equation is easier to process:

0b_0000_0011 AND 0b_0000_0001 = 0b_0000_0001

If you compare the two numbers bit by bit using the truth table the result is 0b_0000_0001, so 1. That means that CVR0 now has value 1. You can perform AND operations like this in the Windows calculator as well by the way.

For CVR1 the result is 0b_0000_0010, but that doesn't fit because CVR1 is a bit and can only be zero or one. To fix that we move the answer one bit - or one position - to the right using the >> operator, the "right shift". The result of that shift is 0b_0000_0001 and that fits exactly. So CVR1 gets value 1.

CVR1 = (sensitivity & 0b_0000_0010) >> 1

For CVR2 the result is 0b_0000_0000. That is shifted right two positions in case bit 2 would be one. The result remains 0b_0000_0000 of course, so CVR2 gets value 0.

CVR2 = (sensitivity & 0b_0000_0100) >> 2

If you do the same for CVR3 you will see that this is also 0. This means that sensitivity 3 has been transferred to CVR3:CVR0 exactly right. The resulting voltage by the way will be 3/24 * 5 = 0.625 volts.

```
-- JAL 2.4j
include 16f877A_bert

-- define the variables
var byte sensitivity
```

```
-- define the pins
pin_d2_direction = output
pin_a0_direction = input
pin_a2_direction = input
pin_a3_direction = input
pin_e0_direction = input

-- set startposition
pin_d2 = low

-- switch on first comparator
cmcon = 1

-- bit 7 CVREN: Comparator voltage Reference Enable bit
CVREN = 1 -- circuit powered on

-- bit 6 CVROE: Comparator VREF Output Enable bit
CVROE = 1 -- voltage level is output on RA2/AN2/VREF-/CVREF pin

-- bit 5 CVRR: Comparator VREF Range Selection bit
-- 1 = 0 to 0.75 CVRSRC, with CVRSRC/24 step size
CVRR = 1

forever loop

   -- measure position of the potmeter
   sensitivity = ADC_read_low_res(5)

   -- adjust reference output accordingly
   sensitivity = sensitivity/17
   CVR0 = (sensitivity &  0b_0000_0001)
   CVR1 = (sensitivity &  0b_0000_0010) >> 1
   CVR2 = (sensitivity &  0b_0000_0100) >> 2
   CVR3 = (sensitivity &  0b_0000_1000) >> 3

   if C1OUT then
      pin_d2 =! pin_d2
      delay_100ms(1)
   end if

end loop
```

You can adjust the sensitivity of this project using the potmeter. If the sound is loud enough the LED will change status, so if it is on it will go off, and vise verse. Since there is no capacitor or coil involved, just a simple LED, you can use this command to flip the pin:

    pin_d2 =! pin_d2

Using an intermediary variable as flag is not necessary.

7.3 Artificial Ears

This project uses two microphones to localize the direction of a sound, just like we do with our two ears. Hence the name of the project: artificial ears.

Technical background

In this project we will use both comparators, one for each ear.

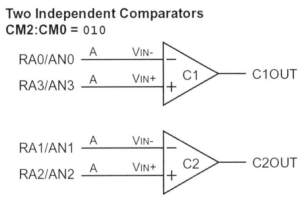

Figure 86. Both comparators in use.

The register settings are identical to the previous project, including the use of the CVR and the sensitivity adjustment technique.

Hardware

Each Electret microphone is connected to a comparator. The V_{IN+} inputs of both comparators are connected to each other. Through the wiper of the potmeter they are connected to the microphones. This way the potmeter can be used to control the balance - or rather the left/right sensitivity of the ears. Instead of a single 33 k resistor you will see a small network on pin 1. The manufacturer of the microcontroller suggests that this is a better way. In practice you will see both techniques. To be honest I have never seen any difference in behavior. But from now on when you see a project with this network you will not be surprised. In the schematic a blue and yellow LED are used, but of course you can pick any color you like

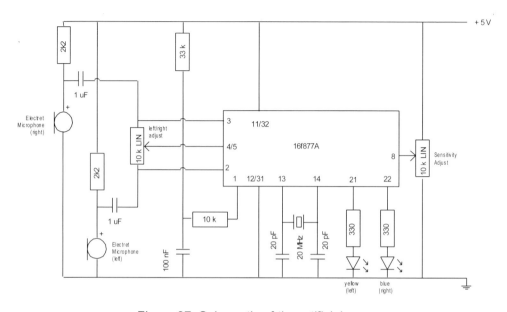

Figure 87. Schematic of the artificial ears.

The next Figure shows the setup in practice. In the foreground you see the two ears (Electret microphones). When a sound moves from left to right past this setup first the left - yellow - LED will light, and somewhere in the middle the blue LED will light. You can use the balance potmeter to adjust the point where one LED goes off and the other goes on to set it right in the middle. Use the sensitivity potmeter to fine-tune the overall sensitivity. Once the potmeters are in the correct position the LEDs will indicate the direction where sounds are coming from.

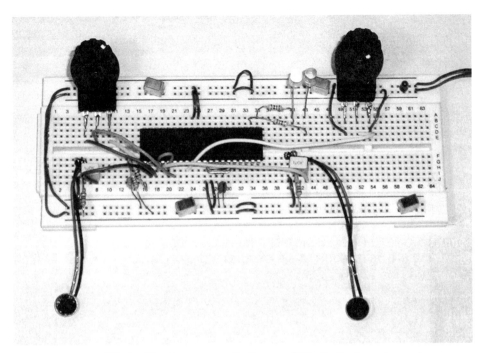

Figure 88. On the foreground both ears (Electret microphones).

Software

The program contains no surprises. The variables c1out and c2out are used to determine the direction of the sound.

```
-- JAL 2.4j
include 16f877A_bert

-- define the variables
var byte sensitivity

-- define the pins
pin_d2_direction = output    -- yellow indicator LED (left)
pin_d3_direction = output    -- blue indicator LED (right)
pin_a0_direction = input     -- MIC left
pin_a1_direction = input     -- MIC right
pin_a2_direction = input     -- voltage input Vref-
pin_a3_direction = input     -- voltage input Vref+
pin_e0_direction = input     -- var resistor 10 kohm
```

```
-- set startposition
pin_d2 = low
pin_d3 = low

-- switch on two common reference comparators
cmcon = 0b_0000_0100

-- bit 7 CVREN: Comparator voltage Reference Enable bit
CVREN = 1 -- circuit powered on

-- bit 6 CVROE: Comparator VREF Output Enable bit
CVROE = 1 -- voltage level is output on RA2/AN2/VREF-/CVREF pin

-- bit 5 CVRR: Comparator VREF Range Selection bit
-- 1 = 0 to 0.75 CVRSRC, with CVRSRC/24 step size
CVRR = 1

forever loop

   -- measure position of the potmeter
   sensitivity = ADC_read_low_res(5)

   -- adjust reference output accordingly
   sensitivity = sensitivity/17
   CVR0 = (sensitivity &  0b_0000_0001)
   CVR1 = (sensitivity &  0b_0000_0010) >> 1
   CVR2 = (sensitivity &  0b_0000_0100) >> 2
   CVR3 = (sensitivity &  0b_0000_1000) >> 3

   -- switch on the proper LED depending on the direction where the
   -- sound is coming from
   if C1OUT > C2OUT then
      -- sound on the left
      pin_d2 = high
      pin_d3 = low
      delay_100ms(1)
   else
      if C1OUT < C2OUT then
         -- sound on the right
         pin_d2 = low
         pin_d3 = high
         delay_100ms(1)
```

```
    else
        -- no sound above sensitivity level
        pin_d2 = low
        pin_d3 = low
    end if
end if

end loop
```

7.4 Frequency meter

In this project we will measure the frequency of a tone using the PWM module of an 16f877A microcontroller.

Technical background

CCP1CON REGISTER/CCP2CON REGISTER (ADDRESS: 17h/1Dh)

U-0	U-0	R/W-0	R/W-0	R/W-0	R/W-0	R/W-0	R/W-0
—	—	CCPxX	CCPxY	CCPxM3	CCPxM2	CCPxM1	CCPxM0
bit 7							bit 0

bit 7-6 **Unimplemented:** Read as '0'

bit 5-4 **CCPxX:CCPxY**: PWM Least Significant bits
<u>Capture mode</u>:
Unused
<u>Compare mode</u>:
Unused
<u>PWM mode</u>:
These bits are the two LSbs of the PWM duty cycle. The eight MSbs are found in CCPRxL.

bit 3-0 **CCPxM3:CCPxM0**: CCPx Mode Select bits
0000 = Capture/Compare/PWM disabled (resets CCPx module)
0100 = Capture mode, every falling edge
0101 = Capture mode, every rising edge
0110 = Capture mode, every 4th rising edge
0111 = Capture mode, every 16th rising edge
1000 = Compare mode, set output on match (CCPxIF bit is set)
1001 = Compare mode, clear output on match (CCPxIF bit is set)
1010 = Compare mode, generate software interrupt on match (CCPxIF bit is set, CCPx pin is unaffected)
1011 = Compare mode, trigger special event (CCPxIF bit is set, CCPx pin is unaffected); CCP1 resets TMR1; CCP2 resets TMR1 and starts an A/D conversion (if A/D module is enabled)
11xx = PWM mode

Figure 89. The ccp1con register of the 16f877A microcontroller.

The PWM module of the 16f877A can, besides generating a PWM signal, also be used to detect changes on a pin. This functionality is called Capture. A change means that the signal suddenly rises - rising edge - or drops - falling edge. When this change occurs a flag is set. For PWM module 1 this flag is called ccp1if, and for module 2 the flag is called ccp2if. In this project we will use module 1.

Detecting a sudden change as an event is nothing special, any pin can do that. But the Capture mode has some extra functionality. Bit 3:0 in the ccp1con register determines to which event the module should react. We don't particularly care whether the rising or falling edge is detected, so we select rising. Next we can choose if we want to detect every edge, every 4th edge or every 16th edge. What we choose depends on the frequency we want to measure. The plan is to run a timer in between two detections, and based on the value of this timer draw conclusions as to the frequency of the detected edges. If the edges are very close together the timer will not have much time to run, making the measurement impossible or inaccurate. In that case it would be a good idea to skip a few edges. For now we select that we want to detect every edge, and get back to this subject later to verify that this is the correct decision. The other bits are not important so the setting of ccp1con is:

    ccp1con = 0b_0000_0101

So far we when we needed a timer we used timer0. The 16f877A has three timers, so let's select timer1 for a change. Setting timer0 was done using the t0con register, so it probably will not surprise you to find out that for timer1 we use the t1con register.

T1CON: TIMER1 CONTROL REGISTER (ADDRESS 10h)

U-0	U-0	R/W-0	R/W-0	R/W-0	R/W-0	R/W-0	R/W-0
—	—	T1CKPS1	T1CKPS0	T1OSCEN	T1SYNC	TMR1CS	TMR1ON
bit 7							bit 0

bit 7-6   Unimplemented: Read as '0'

bit 5-4   T1CKPS1:T1CKPS0: Timer1 Input Clock Prescale Select bits
11 = 1:8 Prescale value
10 = 1:4 Prescale value
01 = 1:2 Prescale value
00 = 1:1 Prescale value

bit 3   T1OSCEN: Timer1 Oscillator Enable Control bit
1 = Oscillator is enabled
0 = Oscillator is shut-off (the oscillator inverter is turned off to eliminate power drain)

bit 2   T1SYNC: Timer1 External Clock Input Synchronization Control bit
When TMR1CS = 1:
1 = Do not synchronize external clock input
0 = Synchronize external clock input
When TMR1CS = 0:
This bit is ignored. Timer1 uses the internal clock when TMR1CS = 0.

bit 1   TMR1CS: Timer1 Clock Source Select bit
1 = External clock from pin RC0/T1OSO/T1CKI (on the rising edge)
0 = Internal clock (FOSC/4)

bit 0   TMR1ON: Timer1 On bit
1 = Enables Timer1
0 = Stops Timer1

Figure 90. The t1con register of the 16f877A microcontroller.

We will not start the timer just yet - we need to wait for an edge detected signal first, so at this moment we are only interested in setting the prescaler. The question is: do we need to slow this counter down using the prescaler. So really the question is: which frequencies do we plan to measure. For the sake of argument we will assume a range between 100 and 20,000 Hz. If you want you can modify this later to other values of your liking.

The crystal has a frequency of 20 MHz. Every command takes 4 clock ticks so the frequency of the commands is 20 MHz / 4 = 5 MHz. This is also the frequency of timer1 - without prescaler that is. This means that one tick takes $2 \cdot 10^{-7}$ seconds. If the sound frequency is 100 Hz the time between two peaks is 1/100 = 0.01 second. During that time timer1 will have counted 50,000 ticks. Timer1 is a 16 bit timer and can count to a maximum of 65,535 so we are ok here. If we want to measure lower sound frequencies we need to enable the prescaler, or not count every single edge.

The same calculation can be made for the upper range of the frequencies we want to measure: 20,000 Hz. At that frequency the time between two peaks is $1/20,000 = 5 \; 10^{-5}$. During that time timer1 has counted 250 ticks. That should be enough for an accurate measurement.

So for the frequency range we want to measure we can react to every edge, and use no prescaler for timer1. For t1con this means we will use the following setting:

    t1con = 0b_0000_0000

The program itself will first have to wait for an edge to occur - the odds of switching the program on exactly at an edge are zero.

    while !ccp1if loop end loop

As soon as the edge has occurred ccp1if will be 1 - true. As long as that hasn't happened the program must wait. When the edge is detected timer1 must be started immediately.

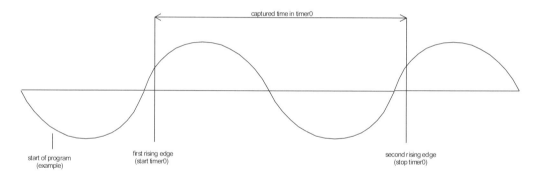

Figure 91. The global overview of program steps.

The program must then wait for the next edge detection, and then stop the timer. The value of timer1 can now be used to calculate the frequency of the original signal using this formula:

$$\text{frequency} = \frac{1}{\text{timer1} * 2 \; 10^{-7}}$$

This calculation will take place in a program on the PC, so you can view the result immediately, and store it in a file is needed for future processing.

Hardware

In theory the sound source can be connected directly to the PWM pin. You will find however that if you try that you need quite a bit of volume. The PWM pin is a digital pin, so you need a minimum of 2 volts before the microcontroller will recognize it as a one - and thus a rising edge. That doesn't sound like much, but for sound that is an impressive volume level. If you want feel free to give it a try.

It is much more convenient to preprocess the sound source using a comparator. Then the voltage only has to be above the level that you can set using a potmeter. If you keep that threshold low you can also use a - relatively - low sound volume. The output of the comparator is connected directly to the PWM pin, so that the PWM unit will only see a decent zero or one. Unfortunately that output is pin a4 which as we know can make itself low but not high. So we need a pull-up resistor to +5 volts.

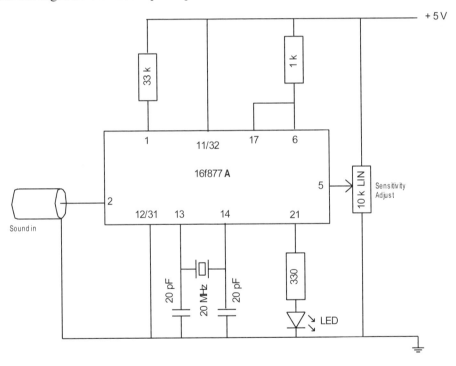

Figure 92. The schematic of the frequency meter.

The wire on the left-hand side goes to the sound source. Make sure to connect the ground (usually the shielding) of the wire to the 0, and the core to pin 2. The wires at the top are from the Wisp648 programmer that is used for the communication to the PC. You can remove the white and yellow wires but all the other must remain in place after programming.

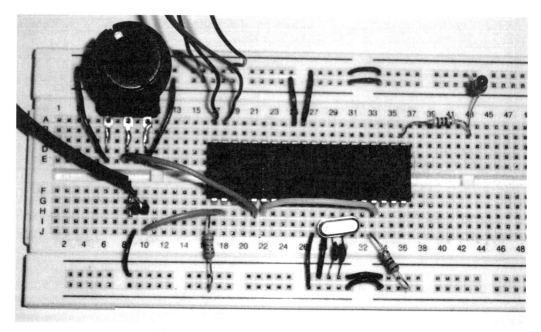

Figure 93. The frequency meter in action.

Software

Timer1 is a 16 bit timer that consists of two 8 bit parts: tmr1h and tmr1l. By combining these you can obtain the 16 bit word using this formula

    word = tmr1l + 256 * tmr1h

Both parts must be cleared.

    tmr1h = 0
    tmr1l = 0

Then the edge detection flag must be cleared.

    ccp1if = 0

And then we wait until the flag is true. If that happens we need to clear the flag and start the timer.

```
while !ccp1if loop end loop
ccp1if = 0
tmr1on = 1
```

Now we wait for the next capture. As soon as that occurs the timer is stopped, both timer values are copied into variables and then send to the PC for further processing.

```
while !ccp1if loop end loop
tmr1on = 0
lowbyte = ccpr1l
highbyte = ccpr1h
serial_sw_write(lowbyte)
delay_10ms(1)
serial_sw_write(highbyte)
```

You may have noticed that there is quite a bit of comment in the source near the register settings. Depending on the frequencies that you want to measure you may want to change the settings of t1con and ccp1con. By copying the possibilities into the program as comments you don't have to consult the datasheet so you can save yourself some time.

```
-- JAL 2.4j
include 16f877A_bert

-- define the variables
var byte highbyte, lowbyte

-- define the pins
pin_d2_direction = output    -- red LED
pin_c1_direction = input     -- capture
pin_a0_direction = input     -- voltage input 1 (from sound source)
pin_a3_direction = input     -- voltage input 2 (var resistor)
pin_a4_direction = output    -- output LED

-- set startposition
pin_d2 = high

-- set CCP1 to capture mode
ccp1con = 0b_0000_0101    -- every rising edge        0b_0000_0101
                          -- every 4th rising edge  0b_0000_0110
                          -- every 16th rising edge 0b_0000_0111

-- set t1con but timer still off, set timer to zero
```

```
t1con = 0b_0000_0000        -- prescaler 1:1 _0000_
                            -- prescaler 1:2 _0001_
                            -- prescaler 1:4 _0010_
                            -- prescaler 1:8 _0011_

-- switch on first comparator
cmcon = 0b_0000_0001

forever loop

   -- clear the counter
   tmr1h = 0
   tmr1l = 0

   -- reset the capture flag
   ccp1if = 0

   -- wait for the first capture to take place
   while !ccp1if loop end loop

   -- now start the timer and clear the capture flag
   ccp1if = 0
   tmr1on = 1

   -- wait for the next capture
   while !ccp1if loop end loop

   -- stop the timer and record the values
   tmr1on = 0
   lowbyte = ccpr1l
   highbyte = ccpr1h

   -- send the data to the PC
   serial_sw_write(lowbyte)
   delay_10ms(1)
   serial_sw_write(highbyte)

   -- change LEDs
   pin_d2 = !pin_d2
   delay_100ms(1)

end loop
```

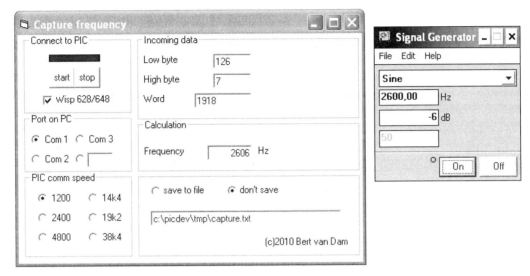

Figure 94. The capture frequency program and the signal generator.

The project works as follows:

1. Connect the hardware to a sound source, in this project a loudspeaker of your PC.
2. Switch the project on.
3. Start the Capture Frequency program that you will find in the free download package, select the correct COM port and click on "Start". The baudrate - 1200 - is preset.
4. Switch the frequency generator - also part of the download package - on, and chose a sinus as signal shape. Set the frequency to for example 2600 Hz and click on the "On" button.
5. Turn the volume of the sound source up quite a bit.
6. Turn the potmeter until the red LED starts flashing. On the PC measurements arrive as low byte and high byte, and are turned into a word. This is the number of timer ticks between two rising edges of the sound signal. Based on this the frequency is calculated and displayed. In the previous Figure you can se that the measurement is quite accurate.

Please note that the calculated frequency is only correct if you use a microcontroller that runs on 20 MHz. That applies for example to the 16f877A but not for the 18f4455 that runs on 48 MHz. The PC program also assumes that you used no prescaler and that the detection takes place at every edge. If not you need to manually calculate the frequency based on the timer ticks on the display. Or write your own PC program. If you own VB5.0 you can adapt the source code, for this is als part of the download package.

7.5 Microphone pre-amplifier

You can use the previous project with a microphone, but you will probably need a small pre-amplifier will make your life much easier. You can built a pre-amplifier using the following schematic. The software is of course identical. Even with this pre-amplifier it is still advisable to use the comparator and thus connect the pre-amplifier to pin 2

Technical background

The design of this microphone pre-amplifier is by Christopher Hunter[29]. It is based on the LM358, an operational amplifier (opamp) specifically design for sound applications The Electret microphone is in the well known network and connected to the non-inverting input of the opamp. The two 100 k resistor set the nominal voltage to 2.5 volts thus ensuring a maximum voltage swing within the range of the microcontroller. The inverting input has a 10 uF capacitor to block DC. The feedback loop ensures an amplification between 46 and 273 times depending on the position of the 500 k potmeter which serves as volume control. The 100 k resistor on the output prevents the microcontroller pin from floating, so it can be directly connected to a microcontroller.

Hardware

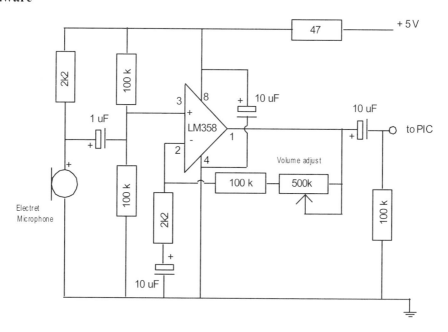

Figure 95. Microphone pre-amplifier.

---

[29] Used with permission.

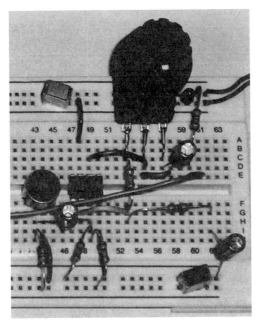

Figure 96. Microphone pre-amplifier.

8 Sensors

In this chapter we will discuss a series of sensors that can be connected to a microcontroller. Sensors are very important because they are the only link between a microcontroller and the outside world. So technically speaking a switch is a sensor too, because it senses... your fingers.

Other places in this book where you will find projects that uses sensors of some sort

Project	Name	Sensor
7.2, 7.3	Sound switch	Microphone
5.2, 5.3, 5.4	Zero crossing	Wire
9.2	VT52 terminal	Potmeter
9.3	IR - receiver	Infrared
10.1 - 10.4	Vision projects	Camera
11.6	Laser alarm	LDR

8.1 Hall effect object protection

The goal of this project is to protect an object without visible wires or contacts.

Technical background

When a current runs through a magnetic field a small voltage is generated perpendicular to that magnetic field. This is called the Hall effect, after researcher Edwin Hall. By measuring that voltage, changes in the magnetic field can be detected. Completely removing the magnet that causes the magnetic field can of course easily be detected by such a sensor. That makes this sensor ideal for non-contact object protection. In this project we use the UGN3140u Hall effect sensor.

Mount the Hall effect sensor at the location where you want to place the object that needs to be protected. Depending on the strength of the magnet the sensor can be half an inch or more from the magnet, and non-magnetic materials - such as a wooden shelf - could be in between. This way both the magnet and the Hall effect sensor and its wires can be hidden from view.

This sensor is direction sensitive and only reacts if the magnetic south pole faces towards the side of the sensor that has the lettering. Keep that in mind when placing sensor and magnet.

Hardware

If you hold the Hall effect sensor in such a way that the lettering is facing you, and the pins are pointing down then the leftmost pin is pin 1 (+ 5 volts), the center pin is pin 2 (0) and the right pin is pin 3 (output). In the breadboard picture the sensor is facing down.

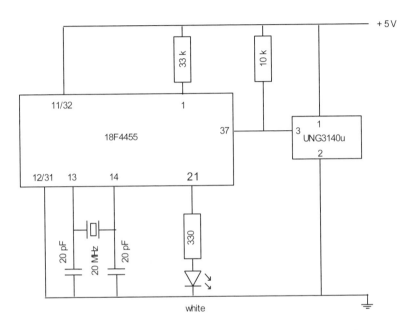

Figure 97. Schematic of the Hall effect object protection.

The magnet on the breadboard is in the correct position to light the LED. If you want more noise than just the lighting of a LED you could replace it by the Monacor MEB-6 buzzer of project 8.4 (low voltage alarm). In that case both the LED as well as the current limiting resistor need to be removed.

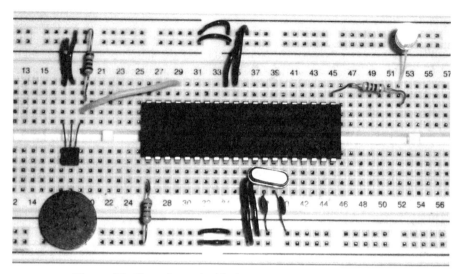

Figure 98. Breadboard with the magnet and the sensor.

Glue the magnet to the bottom of the object that you want to protect. In this case a beautiful Chinese Cloisonné vase.

Figure 99. Chinese vase with the magnet glued to the bottom.

Software

The Hall effect sensor is connected to pin b4. When the sensor no longer sees the magnetic filed - apparently the object is removed - this pin will go high. The program uses the variable "alarm" to remember if the LED has already been on for 10 seconds continuously. At the start of the program the variable is false. If the object has disappeared pin_b4 is high so this condition is true:

   if pin_b4 & !alarm then

Once the LED has been on for 10 seconds the variable alarm is made true, which causes this condition to fail. From now on the LED will just flash until the object has been replaced, at which case the project is reset.

```
-- JAL 2.4j
include 18f4455_bert

-- define pins
pin_b4_direction = input
pin_d2_direction= output

-- LED off
pin_d2 = low

-- declare variable
var bit alarm = false

forever loop

  -- magnet removed?
  if pin_b4 & !alarm then
    -- light alarm LED for 10 seconds and remember
    -- that the alarm is activated
    alarm = true
    pin_d2 = high
    delay_1s(10)
    pin_d2 = low
  end if

  if !pin_b4 then
    -- has the object been replaced
    alarm = false
```

```
    pin_d2 = low
  end if

  if alarm then
    -- is the alarm still activated then
    -- flash the alarm LED
    pin_d2 = !pin_d2
    delay_100ms(5)
  end if

  end loop
```

8.2 Touch key

In the book PIC Microcontrollers[30] a human sensor is described that uses a small aluminum sheet as a single sensor contact. This sensor reacts to humans as soon as they get near it. In this project we will use the same principle to make a touch key. It uses two contacts which allows static electricity to be safely shortened to ground. In theory anyway.

Technical background

The technique is very interesting. Two pins of the microcontroller are connected to each other with a large resistor, of for example 470 k ohm. One of the pins is also connected to a metal contact (on the breadboard this is the wire of the resistor). Close to this contact, without touching it., is a second contact connected to the ground. Between the two contacts is about 2 mm of space. The program makes pin 21 high and waits a bit.

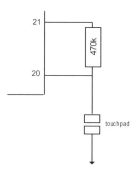

Figure 100. Touch key sensor.

---

[30] PIC Microcontrollers, 50 projects for beginners and experts. Originally based on an article in Byte Craft Limited, used with kind permission.

Next the program checks to see if pin 20 is high. If this is not the case it can only be caused by a connection between the two sensor contacts, for example because someone touches it with a finger. The LED is then switched on for 1 second. If pin 20 is high then there apparently is no finger so the LED is not switched on, and the process repeats itself.

Hardware

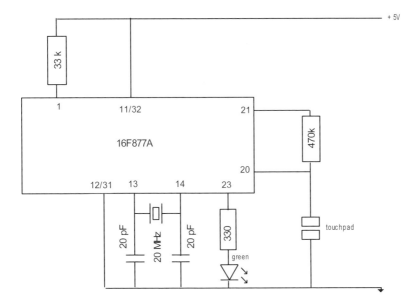

Figure 101. Schematic of the touch key.

Microcontrollers can easily be damaged by static electricity. It is probably a good idea to shape the ground contact in such a way that the finger touches that contact first before it touches the one connected to the microcontroller pin, thus diverting static electricity as much as possible. The risk of damaging the microcontroller can however never be completely avoided.

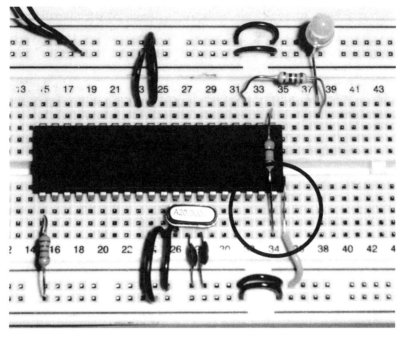

Figure 102. The project on the breadboard with a circle around the touch key.

Software

The program executes the following steps:

1. Make pin 21 low.
2. Wait a bit.
3. Make pin 21 high.
4. Wait 4 to 5 microseconds.
5. See if pin 20 is high.
6. Take the appropriate action.

If pin p20 is still low the LED must be switched on for one second.

```
-- JAL 2.4j
include 16f877a_bert

-- define the pins
pin_c4_direction = output
pin_d2_direction = output
pin_d1_direction = input
```

```
--define variable
var bit led is pin_c4
var bit sensorout is pin_d2
var bit sensorin is pin_d1

forever loop

    -- discharge the sensor
    sensorout = low
    delay_1ms(1)

    -- charge the sensor
    sensorout = high
    delay_4us

    -- get the status
    if sensorin then
       led = low
    else
       led = high
       delay_1s(1)
    end if

end loop
```

8.3 Capacitive (no contact) level gauge

In this project we will measure the liquid level in a plastic cup without bringing a sensor in contact with the liquid. This way the level of a corrosive liquid that otherwise would chemically attack the sensor, can safely be measured from the outside of a vessel.

As it turns out this sensor can also be used as a "someone has his finger in the water" sensor. Although the practical purpose of such a sensor is not clear right off the bat.

Technical background

We use the same technique as the previous project. This time however the contacts aren't short-circuited by a finger, but they are the "plates" of a capacitor, glued to two sides of a plastic cup. The capacitor is made up with two strips of metal - aluminum foil in my case - and taped the to two sides of the cup (on the outside) and the content of the cup. One strip is connected to the ground, the other to the microcontroller. The metal strips of course are only on the sides and do not touch! The cup is now basically a large capacitor with the content of the cup - air at this point - as dielectric. If you fill the cup with water than this will replace the air, thus changing the dielectric, and thus the value of the capacitor. The value of a capacitor can be calculated using this formula:

$$C = \frac{e * A}{d}$$

Where:
C = value of the capacitor
e = dielectric constant
$e_{water} = 80$
$e_{air} = 1$
A = surface of the plates
d = distance of the plates

As you see there is a huge difference between the dielectric constants of air and water (a factor 80) so the difference should be quite obvious. In principle we should be able to calculate this, but it is easier to measure it using the software at the end of this project. We will make pin 21 high, and measure the time it takes for pin 20 to become high too. How long this takes depends on the value of the capacitor, for it has to be loaded first. The higher the water level, the higher the average dielectric constant, the higher the value of the capacitor, the longer it takes to charge it. So there must be a direct relationship between the water level and the charge time.

water level (cm)	charge time (uS)
0	2.5
1	3.5
2	5.0
3	6.0
4	7.0
5	8.0
6	9.5
7	11.0

Table 26. Relationship between water level and charge time.

The cup is not 100% pipe-shaped but the relationship should be more or less linear, and we can see in the graph that this is indeed the case. A charge time such as 2.5 is measured because the value was fluctuating between 2 and 3.

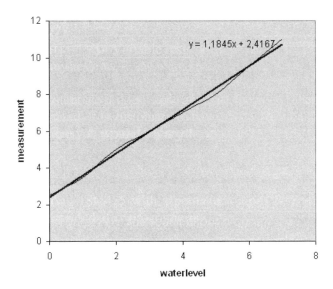

Figure 103. Relationship between water level and charge time.

This means that the following formula applies:

charge time = 1.18 * water level + 2.42

Or in a more useful format, so you can determine the water level based on the measured charge time:

$$\text{water level} = \frac{\text{charge time} - 2.42}{1.18} \quad (cm)$$

Interestingly enough you can use this technique also to see if there is just water in the cup, or if someone is holding a finger in the water. This finger will change the dielectric as well, and that will influence the measurement. The next Figure shows the charge time in MICterm, and the moment that a finger is inserted in the water is clearly visible. So this sensor is also a "finger in the water" sensor.

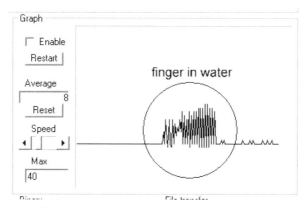

Figure 104. Effect of a finger in the water.

Hardware

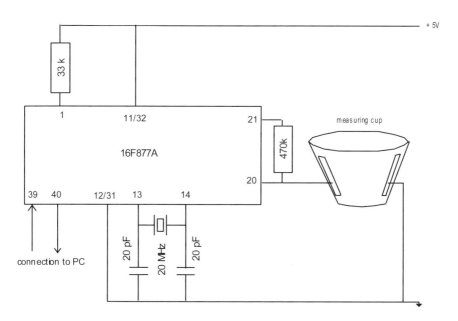

Figure 105. Schematic of the water level meter.

The wires are soldered (with abundant solder - aluminum foil is hard to solder) to the aluminum foil strips and then wires and foil are taped to outside the cup. Note that these are two separate strips, the do not touch, nor do the wires. They are on the outside of the cup so they do not touch the liquid either.

Figure 106. The project in real life.

Software

The charge time is send to the PC, in uS. This value is undoubtedly incorrect because of the loop overhead in the program, but as a relative measurement it has proven to be very useful. You can use this technique also to make a level alarm. Because the measurement is varying quite a bit you should use an average value for that.

In the picture the cup - filled with water - is close to the microcontroller for esthetic reasons. In real life I would suggest a larger distance since it is probably not a good idea to get water in your breadboard or on your microcontroller.

```
-- JAL 2.4j
include 16f877a_bert

-- define the pins
pin_c4_direction = output
pin_d2_direction = output
pin_d1_direction = input

--define variable
var bit led is pin_c4
```

```
    var bit sensorout is pin_d2
    var bit sensorin is pin_d1
    var byte counter = 0

    forever loop

        -- discharge the sensor
        sensorout = low
        delay_1ms(1)

        -- charge the sensor
        sensorout = high

        -- measure the charge time
        while !sensorin loop
          counter = counter+1
          delay_1us
        end loop

        -- send charge time to the PC
        serial_sw_write(counter)
        delay_100ms(1)
        counter = 0

    end loop
```

8.4 Low voltage alarm

In this project we will built an alarm that reacts to a low voltage. This is a special project because the microcontroller itself is using that same (low) voltage.

Technical background

A microcontroller can measure the voltage of a pin using the A/D module. In this module the voltage is compared to a reference voltage, normally speaking the voltage that powers the microcontroller, also known as V_{DD}. If the voltage on the pin drops then so does the measured value, because the voltage drops relative to V_{DD}. In this project the entire microcontroller is running off the same voltage that we want to keep an eye on. For that reason the A/D measurement would always yield the same result, regardless of the actual voltage.

The solution is to use a reference voltage. This is an electronic component that always generates the same voltage regardless of the voltage of its power supply. Assuming of course that the power supply voltage is larger than the reference voltage.

This means we can use the A/D unit, however the effect will be the reverse of what you might expect. If the measurement increases this means that the power supply voltage, or V_{DD}, decreases, for the voltage on the A/D pin doesn't change.

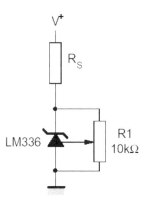

Figure 107. Basic connection of the LM336.

We will use the LM336 voltage reference. It must be powered through a resistor Rs. How large this resistor must be can be calculated based on the amount of current that the unit can handle. According to the datasheet the current should be between 400 uA and 10 mA. As the voltage V^+, which is of course the same as V_{DD}, decreases the current also decreases. So we will start the calculation with 5 volts and a maximum current of 10 mA. The LM336 regulates itself to 2.5 volts, so at 5 volts that leaves 5 - 2.5 = 2.5 volts for the resistor. To get 10 mA of current at that voltage the resistor should be:

$$R = V / I = \frac{2.5}{10 \cdot 10^{-3}} = 250 \text{ ohm.}$$

This is the minimal resistance. The lowest practical voltage of V^+ is 2 volts, because below that the microcontroller will shut down. In order for the unit to still work the current should be at least 400 uA (400 10^{-6} A).

$$R = V / I = \frac{2}{400 \cdot 10^{-6}} = 5 \text{ k.}$$

That is the maximum resistance. We select a value in between the minimum and maximum: 1 k.

As an alarm we use the Monacor MEB-6 buzzer designed for 3 - 9 V DC, with a power consumption of 25 mA at 6 V. The measured current is 10 mA at 5 volts. That means it can be connected directly to the pin of the microcontroller. The datasheet is quite unclear about it, but it seems logical to assume that the unit contains a coil of some sort. So for safety we will use a protection diode. The buzzer works to about 2.5 volts minimum.

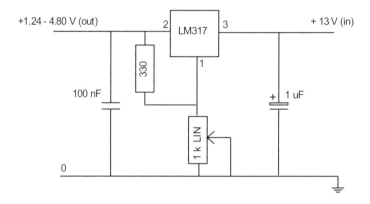

Figure 108. Variable power supply.

Because we want to reduce the voltage we cannot make use of the power supply that is built in the Wisp648. So we will make our own power supply, based on the LM317 voltage regulator. This unit regulates the voltage to 1.2 volts. We can raise this by making the unit regulate the upper half of a voltage divider.

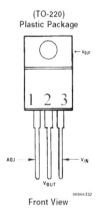

Figure 109. Pain layout of the LM317.

If we select a voltage divider consisting of a fixed 330 ohm and a 1 k LIN potmeter as shown in the previous Figure than the range is 1.2 to 4.8 volts[31]. The range is deliberately chosen very small so you cannot accidentally put too high a voltage on your project, thus damaging the microcontroller

Hardware

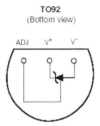

Figure 110. Pin layout of the LM336.

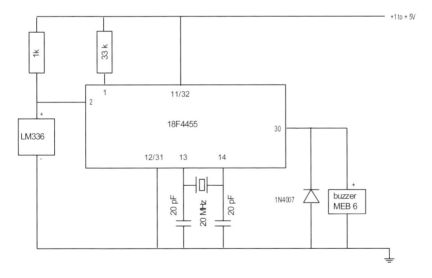

Figure 111. Low voltage alarm (without power supply).

---

[31] For the mathematically inclined readers: over the 330 ohm resistor the voltage is (fixed) 1.2 volts, so the current is $I = V/R = 1.2/330 = 3.64 \cdot 10^{-3}$ A. The potmeter is 1k maximum, at which point the current results in a voltage of $V = I*R = 3.64 \cdot 10^{-3} * 1000 = 3.6$ V. So the total voltage is $1.2 + 3.6 = 4.8$ volts.

The schematic doesn't contain the variable power supply, which has been shown previously. The breadboard does show it, on the far right hand side. Note that pin 1 of the LM336 is not in use.

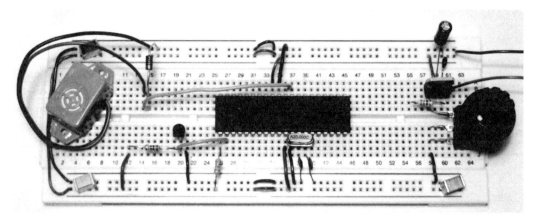

Figure 112. Low voltage alarm with variable power supply.

Attention: if you use the variable power supply you cannot use the Wisp648 power supply at the same time! You must connect the red and black wires, but you must disconnect the power plug from the side of the Wisp648. You can only program the microcontroller if the adjustable power supply is turned up all the way, otherwise the Wisp cannot get enough power to force the microcontroller in programming mode.

Software

The A/D value of pin 2 is determined with the command

    measurement = adc_read_low_res(0)

Normally when you measure a voltage you use the formula $V_{in}/V_{DD} * 255$ to get the result of an 8 bit A/D measurement, where V_{in} is the variable and V_{DD} is fixed (at 5 volts). In this case we do it the same way, however now V_{in} is fixed (at 2.5 volts) and V_{DD} is the variable

Let's assume that the alarm needs to go off at a voltage of 2.7 volts. The measurement that goes with that voltage is:

    $V_{in}/V_{DD} * 255 = 2.5/2.7 * 255 = 236$

If we use this value in the program as threshold the alarm will go off at about 2.7 volts. Note that the buzzer itself will stop at about 2.5 volts, and the microcontroller at about 2 volts.

```
if voltage > threshold then
   pin_d7 = high
else
   pin_d7 = low
end if
```

Putting it all together results in this program:

```
-- JAL 2.4i
include 18f4455_bert

-- define the pins
pin_a0_direction = input       -- voltage input
pin_d7_direction = output      -- output LED

-- set startposition
pin_d7 = low

-- variables
var byte voltage, threshold
threshold = 236                -- 2.7 volts

forever loop

   -- sample ADC
   voltage = adc_read_low_res(0)

   -- compare to threshold
   if voltage > threshold then
      -- sound alarm while below threshold
      pin_d7 = high
   else
      pin_d7 = low
   end if

end loop
```

Make sure that the power supply plug is removed from the Wisp648, so that the Wisp is powered through the breadboard and its variable power supply.

Connect the power to the breadboard, and turn the potmeter all the way to the right (maximum resistance). This is the only way to program the microcontroller. Enter a threshold to your liking into the program and put the program into the microcontroller. Slowly turn the potmeter to the left until the low level voltage is reached and the buzzer sounds.

8.5 Temperature control

In this project we will measure the temperature using an LM35. If the temperature is below a certain threshold a LED will be lit.

Technical background

This sensor, the LM35, generates a voltage that is directly proportional to the temperature of the unit. That means that regardless of the supply voltage 1 degree Centigrade equals 10 mV. So if the voltage is for example 0.172 volts then the temperature is 17.2 degrees.

The nice thing about this sensor is that you do not need to calibrate it, and that it is very easy to use, particularly compared to the sensor in the next project. The downside is that the sensor needs to be relatively close to the microcontroller to prevent any kind of loss in the cabling.

Hardware

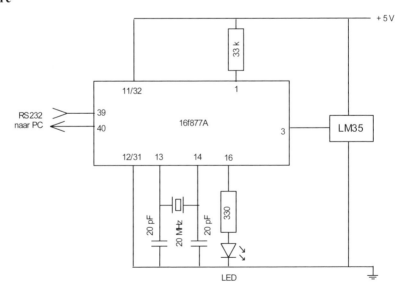

Figure 113. Temperature alarm schematic.

The temperature can be displayed on the PC via the RS232 connection. In that case the Wisp programmer is an integral part of this project and must not be removed after programming.

Figure 114. Temperature alarm in action: low temperature warning LED is on.

Software

The variable "threshold" is the alarm value. In this project the alarm is set on 16 degrees Centigrade, so the warning LED will go on if the temperature drops below that threshold. To prevent flickering of the LED it will not go off again until the temperature has risen above the threshold, rather than on the threshold itself.

```
if temperature < threshold then
    pin_c1 = high
end if
if temperature > threshold then
    pin_c1 = low
end if
```

This time the A/D measurement is executed in high resolution. That means that the maximum measurement is not 255 but 1023 (10 bits). That doesn't fit in a byte, so the variable "resist" is defined as a word.

```
resist = ADC_read (1)
```

The calculation is executed on the right side of the equation, after which the answer is stored in the left side of the equation, in the variable "temperature". On the right side the largest variable type is a word, so the entire calculation is executed in words. For that reason we are allowed to multiply by 100 and by 5 without getting in trouble.

```
temperature = (resist * 100 * 5) / 1023
```

This is why: a word can hold a maximum value of 65,535. That means that resist may not be larger than 65,535/100/5 = 131. A that moment the voltage is 131/1023 * 5 volts = 0.64 volts. One degree is 10 mV so this is equal to 0.64/10 mV = 64 degrees C.

The answer of our temperature calculation is stored in the variable temperature. This is a byte, so it can hold a maximum of 255, which is not a problem for we cannot measure temperatures over 64 degrees anyway.

This is the complete program:

```
-- JAL 2.4i
include 16f877a_bert

-- LED
pin_c1_direction = output

-- variables
var byte temperature, threshold = 16
var word resist

forever loop

   -- convert analog on a1 to digital
   resist = ADC_read(1)

   -- convert to temperature
   temperature = (resist * 100 * 5) / 1023

   -- send data to PC
   serial_sw_write(temperature)
   delay_100ms(1)

   -- if below threshold light the LED
   if temperature < threshold then
      pin_c1 = high
```

```
    end if
    if temperature > threshold then
       pin_c1 = low
    end if

 end loop
```

The measured temperature is send to the PC over an RS232 connection using the Wisp programmer. That means that you can use this project also as a thermometer. The temperature can be shown in a graph in MICterm. If you set the maximum to 25 the vertical axis is used nicely. Or 64 if you want to see the full range.

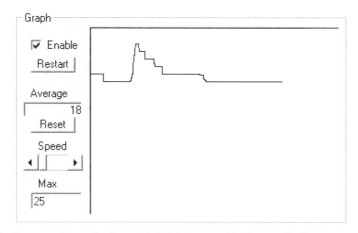

Figure 115. Temperature displayed in MICterm, with the vertical max. on 25.

8.6 Temperature in a poultry farm

Jean Marchaudon, owner of a big poultry farm in France uses JAL to measure and control the temperatures in the different areas of his farm.

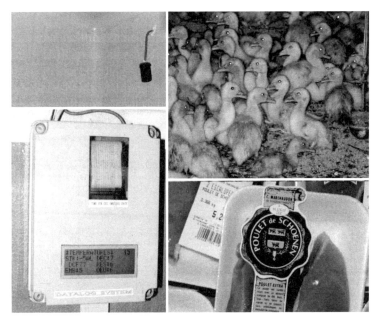

Figure 116. Poultry farm photo collage.

The sensors are mounted in empty film canisters (top left). The measurements are collected in a separate room that controls the heating. All measurements are shown on a display and printed for future reference (left bottom). On the right side the chicks enjoying the heating, and at the bottom you see the final product in the shops.

Technical background

The communication technique used in this project is 1-wire, where all communications run over a single wire. All units are connected to ground, but many types do not even need to be connected to +5 volts. They get their power from the data connection. Because multiple units can be connected to the same wires (or rather: wire) this is an actual bus. The microcontroller controls the communication and is the master, all other units are slaves. The slaves respond to commands of the master, they can never send a message of their own accord.

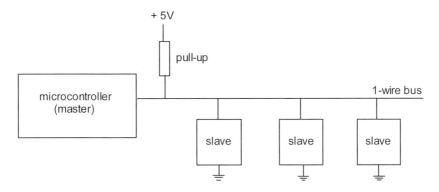

Figure 117. Structure of the 1-wire bus.

Every slave has a unique 64 bit number. This identification number, ID, is entered during manufacturing and cannot be changed later. The ID number functions as an address, when the master wants to address a particular slave it sends a message to that slave's ID number. Only the slave with that ID number will respond.

Communication takes place by making the bus low. A pull-up resistor pulls the bus back up. The bus is very robust and suitable for communication over long distances in (electrically) noisy environments. According to the manufacturer a bus length of up to 300 meters is possible.

Before we continue with this project a short interlude. Take a close look at this program, in combination with the hardware shown in the next two Figures. Do you think this program does anything, and if yes: what?

```
-- JAL 2.4j
include 18f4455_bert

pin_d1_direction = output
pin_d1 = low

forever loop
   pin_d1_direction = input
   delay_1s( 1 )
   pin_d1_direction = output
   delay_100ms( 5 )
end loop
```

When you have made up your mind built the hardware and test the program. Do not continue with this chapter until you have tried the next program and have seen what happens.

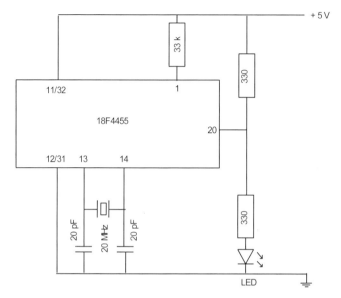

Figure 118. Schematic of the direction test.

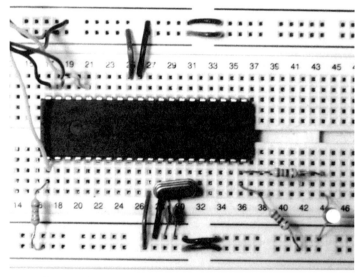

Figure 119. The direction test on the breadboard.

Did you predict correctly what would happen? It seems like this program doesn't do anything useful. Pin d1 is defined as an output, made low, and then in a loop the direction of the pin is changed. That this would make the LED flash is not at all obvious!

So how is this possible? Before the loop starts pin d1 is made an output and made low. Both sides of the LED are now connected to the ground, so it is off. Next - in the loop - the pin is made an input. The microcontroller "let's go" of the pin and checks to see whether the pin is a zero or a one. It will not do anything with this information because you didn't ask it to. But now that it "let go" of the pin, it is no longer low. Through the other resistor a current can flow from the +5 volts that this resistor is connected to, past the pin, through the current limiting resistor of the LED and then through the LED itself to the ground on the other side. And that means that the LED will go on. When the pin is made an output again the microcontroller immediately makes it low again, because that was the last pin related command it received. And that causes the LED to go off again.

We spend a lot of time on this because it explains how 1-wire communication works, which is otherwise almost impossible to grasp. Let's take a look at a small fragment from the library that takes care of 1-wire communication, written by Vasile Surducan:

```
procedure d1w_write_bit( bit in x ) is
   d1w_bus_out = low
   delay_10us( 1 )
   if x == high then d1w_bus_in = high end if
   delay_10us( 8 )
   d1w_bus_out = high
   delay_10us( 1 )
end procedure
```

In this procedure d1w_bus_in is pin d1 itself, and d1w_bus_out is the direction (input or output) of that pin d1. It is important to realize that high for a direction is the same as input and low is the same as output. With a bit of work you can see that if x is high - meaning a 1 must be transmitted - the following takes place on the 1-wire bus:

> low - high - high.

When x is low - a 0 needs to be transmitted - this happens:

> low - low - high

What is so elegant about this construction is that you can listen on the same pin, using this procedure:

```
procedure d1w_read_bit( bit out x ) is
  x = high
  d1w_bus_out = low
  delay_10us( 1 )
  d1w_bus_out = high
  delay_10us( 1 )
  if d1w_bus_in == low then x = low end if
  delay_10us( 7 )
end procedure
```

First it makes the answer 1 (x = high). Then the following signal is put on the bus

low - high

And then the microcontroller waits for an answer. This is possible because high is an input. If the bus remains high, then the answer is one. But if the bus is immediately pulled low - by the slave - then the answer is zero.

In order to use a 1-wire library you don't need to understand how it works. But this is a very special technique that I like very much, so I wanted to share it with you. Ok, let's continue with our project.

In this project we will use the DS1822, a 1-wire temperature sensor made by Dallas Semiconductors. This sensor does require a +5 volts connection.

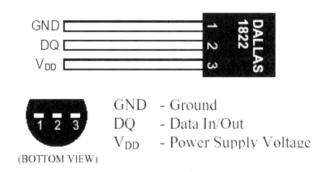

Figure 120. Pin layout of the DS1822.

The first step you need to take is to find out what the ID number is of the component you want to use, because the ID number is that component's address.

The next step is to ask that ID number for the temperature. The slave will answer with the temperature and a control number. If the control number is incorrect there is something wrong with the communication of with the slave. This is important, because if the slave malfunctions, and doesn't reply at all, then the protocol will see this as a series of one's, as you have seen in the intermezzo. If the control number is correct then the temperature is correct too.

The temperature can be above or below zero. Below zero is a problem because microcontrollers cannot calculate with negative numbers by default (although JAL can). The technique used to convey negative numbers is called two's complement. This is the number that you get when the temperature is deducted from a very large power of two. Together they are that big number, so they complement each other. That sounds rather complicated, but it is in fact very simple.

If the temperature is larger than 2048 you take the exact opposite of all bits, and then add one. The result is the negative temperature. Let's look at for example 4041, binary 0b_1111_1100_1001.

	111111001001
flip all bits	000000110110
add one	000000000001
the result	000000110111 = - 55

For the DS1822 it is a bit more complicated because its answer has 16 bits, where the lowest four are decimal digits. We will ignore those, so we divide the answer by 16 to get rid of them. Now we have 12 bits left over. A word has room for 16 bits, so we have 4 unused bits. As such not a problem because they are all zero, but they will be a problem when we flip all bits because then they will be ones. We fix that by erasing them after flipping. That results in this routine (division by 16 has already taken place):

```
if temp > 2047 then
   -- signbits engaged, so negative value
   -- convert using two's a complement
   serial_sw_write( "-" )
   temp = ! temp
   temp = temp & 0b0000111111111111
   temp = temp + 1
end if
```

Hardware

As pull-up resistor for the bus we have chosen a value of 4k7 ohm (4700 ohm). It is sometimes difficult to find a good value, particularly in long busses.

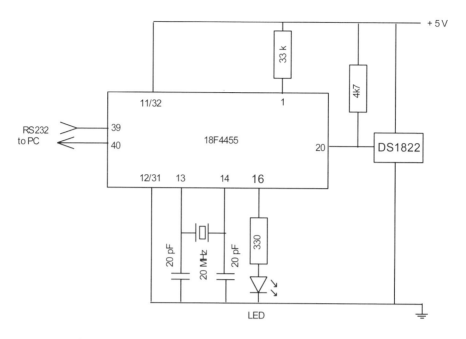

Figure 121. Schematic of the 1-wire temperature meter.

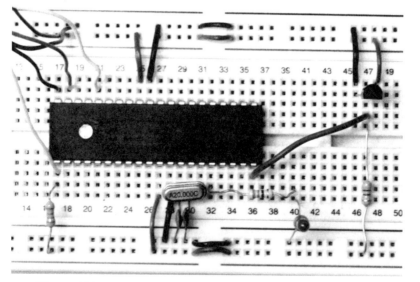

Figure 122. The 1-wire temperature meter on a breadboard.

Software

In order to use 1-wire you need to load a special library. In order to use the DS1882 you also need to load a special library. The order in which you do this is important. The DS1822 library uses commands from the 1-wire library, so the latter must be loaded first. Both libraries use the 18f4455_bert library, which always must be the very first library to load. In fact it should be the very first command, except for comments.

command	description
read_ID	Request the ID number - or address - of a component (slave). There must be only one slave on the bus otherwise they will all reply at the same time.
MatchRom	Attention slaves: the master is about to send an address.
Send_ID	Send the address.
DS1822_start_temperature_conversion	Order the DS1822 to determine the temperature.
DS1822_read_temperature_raw	Read the temperature without checking the CRC (control number). Not recommended.
d1w_read_byte_with_CRC	Read a group of bytes and check the CRC (control number). In order to get the temperature you must read 9 bytes.
Load_My_ID	You can store addresses in the library and retrieve them as needed.

Table 27. Commands in the DS1822 1-wire library.

The next commands are in the 1-wire library. Normally speaking you only need these commands when you are making your own 1-wire component library.

command	description
d1w_write_bit	Write an individual bit.
d1w_read_bit	Read an individual bit.
d1w_write_byte	Write a byte.
d1w_read_byte	Read a byte.
d1w_reset	Give a reset.
d1w_present	See if anyone is present on the bus.

Table 28. Commands from the 1-wire library.

Putting it all together results in this program:

```
-- JAL 2.4i

include 18f4455_bert
include 1_wire
include ds1822_1_wire

-- indicator LED
var bit led        is pin_c1
pin_c1_direction = output

-- read the ID of the DS1822
read_ID
delay_100ms(1)

forever loop

   -- flash the LED so we know the program is running
   led = on
   delay_100ms(1)
   led = off
   delay_100ms(1)

   -- for all probes
   DS1822_start_temperature_conversion

   -- get the temperature from the probe
   MatchRom
   Send_ID
```

```
    d1w_write_byte( 0xBE )
    d1w_read_byte_with_CRC( 9 )

    -- show the result
    if GOOD_crc == 0 then
       -- convert to temperature (table 2, page 4, datasheet ds1822)
       temp = (word(d1)+(256 * word(d2)))/16
       if temp > 2047 then
          -- signbits engaged, so negative value
          -- convert using two's a complement
          serial_sw_write( "-" )
          temp = ! temp
          temp = temp & 0b0000111111111111
          temp = temp + 1
       end if
       format_byte_dec(serial_sw_data,temp,5,0)
       serial_sw_write( "°" )
       serial_sw_write( "C" )
    else
       -- oops, CRC is not correct
       serial_sw_write("C")
       serial_sw_write("R")
       serial_sw_write("C")
       serial_sw_write(" ")
       serial_sw_write("e")
       serial_sw_write("r")
       serial_sw_write("r")
       serial_sw_write("o")
       serial_sw_write("r")
    end if

    -- newline and carriage return (Windows style)
    serial_sw_write( 10 )
    serial_sw_write( 13 )
    delay_1s(1)
 end loop
```

To get the temperature lines neatly under each other you cannot use MICterm, so we will use HyperTerm instead. Unfortunately HyperTerm is not capable of switching the Wisp into passthrough mode. In the download package you will find a small program called Wisppassthrough. This program has a tiny window that is always on top, almost like an extra button in your terminal program. Click on Enable to put the Wisp in passthrough mode. The indicator will turn green. If it stays yellow then the Wisp is not switched on or a terminal packet is using the port. Once the indicator is green you can click on the little telephone in HyperTerm to start the connection. Every time when the Wisp has been switched off, or used for programming, you need to (re-)enable passthrough. If you switch the Wisp off the indicator will remain green because the program doesn't know that you just did that. You can terminate the passthrough mode by clicking on break, but that is not strictly necessary. When you start programming the Wisp automatically terminates the passthrough mode.

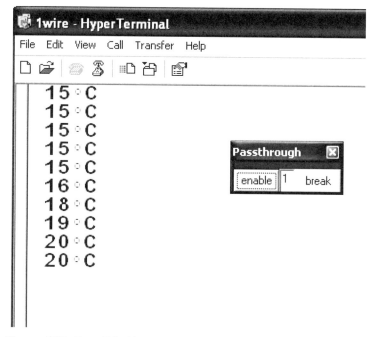

Figure 123. Result in HyperTerm with the Passthrough program.

9 Communication

In this chapter we will discuss projects with several different communication techniques. Communication is important because it is the only way to exchange information with components, PC's or other microcontrollers. In these projects we are not dealing with user communication.

Another place in this book where communication plays an important role:

Project	Name	Communication
8.5	Poultry farm	1-wire

9.1 RS232 - Passthrough communication

RS232 is one of the most often used means of communication between a microcontroller and a PC. While portable PC's with an RS232 port (also known as serial port) are becoming scarce desktop machines still have them. They are also often used in the industry. Small PC's such as portables usually have a USB port. This is discussed later in this chapter.

Figure 124. Wisp in passthrough mode.

One of the special properties of the Wisp programmer is that it is capable of making a passthrough connection between the microcontroller and the PC. The programmer is connected to the programming pins of the microcontroller, and those are never the same pins as the RS232 pins (assuming the microcontroller even has an RS232 unit). For that reason all _bert libraries load a library which enables software RS232 functionality[32]. Since it is software you can have this functionality on any pin that you want, but it is of course most convenient to select the programmer pins, so you don't have to swap any wires. The _bert libraries take care of this by default.

command	description
Serial_SW_Read(data)	Receive data and put it in the variable "data".
Serial_SW_Write (data)	Send the contents of the variable "data".
Serial_SW_Locate(horizontal, vertical)	Move the cursor (on a VT52 terminal or emulation) to the coordinates horizontal, vertical.
Serial_SW_Clear	Clear the screen from the current cursor position (on a VT52 terminal or emulation).
Serial_SW_Home	Move the cursor to the home position - the upper left corner (on a VT52 terminal or emulation).
Serial_SW_Byte(data)	Send a byte named "data" as three digits - or less - instead of a single number.
Serial_SW_Printf(array)	Send a complete array - which needs to be defined first - with a single command. For example const byte mystr[] = "Bert van Dam" followed by Serial_SW_Printf(mystr).

Table 29. Some commands from the serial software library.

Contrary to hardware serial communication the software communication doesn't have any buffers. So when data is coming in when the microcontroller isn't waiting for it, it will be lost. When you are setting up RS232 communication you have to keep this in mind. The previous table shows some of the serial commands that you can use. You will find the complete list in the appendix.

---

[32] With the exception of the 10f200_bert because it doesn't have enough memory.

Once programming is completed the passthrough function is not enabled automatically. The terminal program MICterm is capable of switching the Wisp in passthrough mode if you select Wisp628/648.

If you use another terminal program you can use the WispPassThrough program. This small program is also part of the download package. Once the program is started you will see a small window that it always on top of any other window.

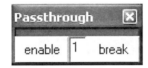

Figure 125. Passthrough window.

If you click on "enable" the Wisp is put in the passthrough mode, and the signal will go from red to green. With "break" you can take the Wisp out of passthrough mode. Normally speaking this is never necessary because when programming starts, or the power is disconnected, the Wisp will disable passthrough mode automatically. Make sure that the COM port is free before enabling passthrough mode. So first enable passthrough and only then connect using the communication package. If the port is not free the signal will go from red to yellow but not to green.

If you want to write your own PC program that switches the Wisp into passthrough mode you can use one of the following techniques.

1. A batch file

All you need is a single line in a batch file:

    c:\picdev2\xwisp\xwisp.exe pass b6t

Should Xwisp be in different location on your PC because you moved it from the default location to another location, you must of course adapt this line.

2. A Visual Basic routine (or any other language)

To get the programmer in passthrough mode the following steps must be taken:

1. Issue a break command for 0.1 second to switch the programmer into attention mode .
2. Use the programmer speed and settings (19200,n,8,1) to issue the passthrough command: 0000p.

3. Wait a bit to give the Wisp time to process that command, for example 0.1 second, or more.
4. Switch back to the PIC baud rate (in the _bert library this is set to 1200 baud).

In Visual Basic (version 5 and 6) this is what that looks like:

```
Private Sub Command1_Click()
    'enable communications
    MSComm1.PortOpen = True

    'switch Wisp programmer to pass-through mode

    'Set the Break condition for 0.1 second to switch the
    'programmer to attention mode
    MSComm1.Break = True
    Duration! = Timer + 0.1
    Do Until Timer > Duration!
        Dummy = DoEvents()
    Loop
    MSComm1.Break = False


    'switch to programmer speed
    MSComm1.Settings = "19200,n,8,1"

    'send pass-through command
    MSComm1.Output = "0000p"

    'wait to make sure the command is processed by the programmer
    Duration! = Timer + 0.1
    Do Until Timer > Duration!
        Dummy = DoEvents()
    Loop

    'switch to microcontroller speed
    MSComm1.Settings = "1200,n,8,1"
End Sub
```

You now have a direct RS232 link to the microcontroller. Obviously a program has to run in the microcontroller that is actually using RS232 in order to communicate with the PC.

Make sure MSComm has the right settings:

> MSComm1.CommPort = 1
> MSComm1.DTREnable = True
> MSComm1.EOFEnable = False
> MSComm1.InputLen = 1
> MSComm1.InputMode = comInputModeText
> MSComm1.NullDiscard = False
> MSComm1.ParityReplace = 0
> MSComm1.RThreshold = 1
> MSComm1.RTSEnable = False
> MSComm1.SThreshold = 0
> MSComm1.Settings = "1200,N,8,1"

Everywhere in this book the speed between microcontroller and PC is set at 1200 baud. That speed could be much higher, 19k2 for example shouldn't be a problem. I use other microcontrollers as well, and some do not have a crystal, which causes problems at higher speeds. Besides the internal doubling of the clock speed of the 18f4455 isn't handled too well by the software library. For that reason I use 1200 baud as the default speed. Of course you can use higher speeds yourself if you want. The speed setting can be found in the _bert libraries.

Establishing communication

In most cases you will be communicating from the PC to the microcontroller using the pass-through function. This means that on the microcontroller you use software serial communication, because the programmer is never connected to the hardware serial pins.

Using software serial communication means that you don't have a buffer available. So data sent to the microcontroller at a moment when it isn't listening is lost. Note that the other way around, from the microcontroller to the PC, is no problem because the PC does have a buffer. This means you have to make sure that the microcontroller is ready and waiting before you send anything. There are three ways to arrange this.

1. You let the user start the communication by having him press a button connected to the microcontroller and then one on the PC keyboard. The microcontroller program makes sure that the Serial_SW_Read(data) command is executed, which means the program will wait for data. Note that the programmer pass-through must be enabled before the user presses the buttons.

2. The microcontroller initiates the communication by sending a signal to the PC, such as the letter "s". The PC does have a buffer and a communication interrupt so it won't miss

any data (assuming the buffer doesn't overflow). As soon as the letter "s" is sent the Serial_SW_Read(data) command is executed so the microcontroller is ready for the PC's reply.

3. You keep track of the communication timing. This is often used in combination with option one or two. If you are sending bulk data to the microcontroller and you know how long it takes to process you simply have the PC send the data at a slightly slower speed.

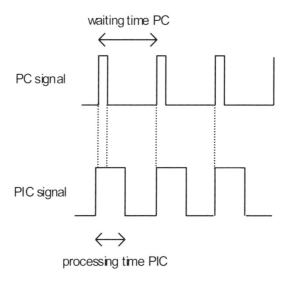

Figure 126 Communication timing loop.

A waiting loop in the PC takes care of the delay. Visual Basic handles that like this:

    Duration! = Timer + 0.1
    Do Until Timer > Duration!
        Dummy = DoEvents()
    Loop

The delay in this case is 0.1 seconds. The "Dummy = DoEvents()" command is a multitasking command. It allows the PC to take care of other things during this loop, such as looking at the mouse position. Without this command the PC will ignore everything (including the mouse and keyboard) while the loop is executing.

9.2 RS232 - VT52 terminal

Many projects in this book make use of MICterm. This program is especially designed for use with microcontrollers and has many qualities that are very useful for example when you are debugging your microcontroller program. Unfortunately you cannot make nice looking screens on your PC using MICterm.

If you use HyperTerm, the free communication package of Windows you can have your microcontroller issue certain commands to get a text at a certain location on the screen[33]. In order to make this possible we will use VT52 terminal emulation. In HyperTerm you can enable this as follows:

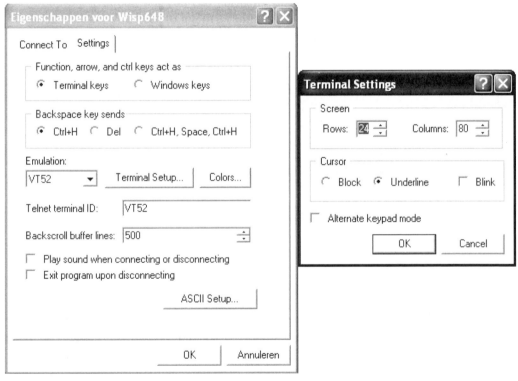

Figure 127. HyperTerm settings for VT52.

---

[33] Windows Vista and 7 do not have this free program anymore. In the download package you will find the free program Hyper Terminal Private Edition (HTPE) manufactured by Hillgreave that has the same functionality. Personally I always use HTPE - on Windows XP - because it has the ability to clear the screen from the menu.

Select port settings:
Bits per second	1200
Databits	8
Parity	none
Stopbits	1
Datatransportcontrol	hardware

You can now use the following commands:

Command	Description
Esc A	Cursor one line up.
Esc B	Cursor one line down.
Esc C	Cursor one position to the right.
Esc D	Cursor one position to the left.
Esc H	Cursor to the home position (top left).
Esc J	Erase everything from the cursor.
Esc K	Erase the rest of the line.
Esc Y row+32 col+32	Go to location row, col.

Table 30. VT52 Escape commands.

Since all commands start with the Escape character they are often called the "escape commands". The ASCII number for Esc is 27. In JAL you can issue the Esc H command as follows:

    serial_sw_write(27)
    serial_sw_write("H")

With the command Esc Y you can move the cursor to a certain location on the screen. For that you need to add 32 to the desired row and column, and then send the ASCII sign that goes with that number.

Let's say you want the text " Variable resistor " to start at row 4 column 3. First you add 32 to these numbers, so you get 36 and 35. In the ASCII table in section 13.4 you check which characters belong to these code. These are $ and #. So the command will be:

    Esc Y$#

Note that there are no spaces between the command and the numbers, so Y, $ and # are against each other. The cursor is now at the correct location, so the text can be printed. In

JAL we do this as follows: first send the Esc character, then send the rest of this instruction with the text that you want to print.

    serial_sw_write(27)
    const byte mystr1[] = "Y$# Variable resistor "
    serial_sw_printf(mystr1)

The text is stored in a constant so it doesn't use any data memory. Then it is printed using the serial_sw_printf command. It is customary to have all declarations at the beginning of the program, so that is where you would declare the string(s) as well.

Hardware

The hardware consists of an 18f4455 with a potmeter on pin 2. The Wisp programmer must of course remain connected after programming.

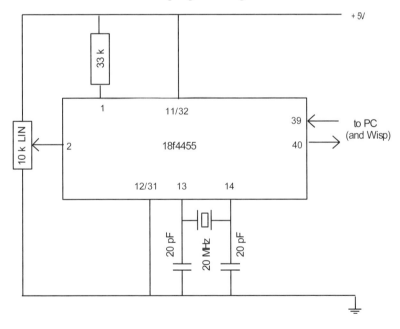

Figure 128. VT52 hardware.

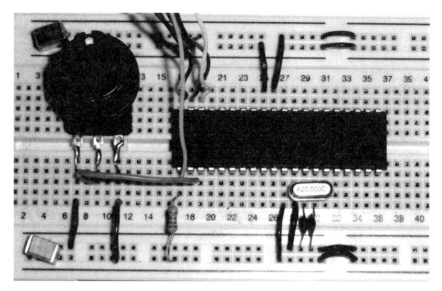

Figure 129. The breadboard, the Wisp is still connected.

Software

The program will print a header, and then the text with the content of the variable "potmeter".

```
--- JAL 2.4j
include 18f4455_bert

-- declare variables
var byte resist
const byte mystr0[] = "VT52 demonstration program "
const byte mystr1[] = "Y$# Variable resistor "
const byte mystr2[] = " units  "

forever loop

  -- print header
  serial_sw_write(27)
  serial_sw_write("H")
  serial_sw_printf(mystr0)

  -- measure voltage on a0
  resist = ADC_read_low_res(0)
```

```
            -- display on VT52 terminal
            serial_sw_write(27)
            serial_sw_printf(mystr1)
            serial_sw_byte(resist)
            serial_sw_printf(mystr2)

            -- wait a bit
            delay_1ms(100)

         end loop
```

Instructions

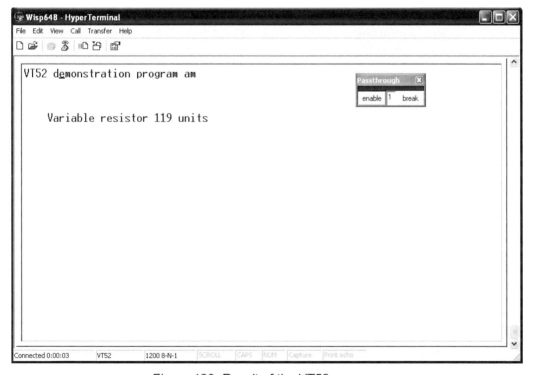

Figure 130. Result of the VT52 program.

1. Power the breadboard and download the program into the microcontroller. Leave the Wisp connected.
2. Start WispPassThrough, enter the correct COM port and click on "enable". The signal will go from red to green.
3. Start HyperTerm.

When you turn the knob you can see the value change on the screen. Without any programming tools on the PC side you can still make very nice looking screens.

Remember to run WispPassThrough before connecting with the terminal program. HyperTerm will auto-connect when it is started. If that happens you will not see anything, so hang up the "telephone" and the click on Enable. If the signal stops at yellow the COM port is still in use. The most likely cause is that HyperTerm is still connected. If the signal remains red you have selected the wrong COM port, or the Wisp is not powered or not connected.

9.3 IR - Receiver

In this project we will make an infrared (IR) receiver that will respond to a Sharp television remote control. In the next project we will make a transmitter that will work with this project, so even if you don't have a Sharp remote control you can still use this project.

Technical background

The first question is what the infrared signal from the remote control looks like. On the internet we find the following information:

```
Sharp

Code length:            17 bits

Carrier:                32kHz

Transmits code + code XOR #7FE0h

header: None

one:     275us pulse + 1900us space
zero:    275us pulse + 775us space

Space between transmissions:    43ms
```

Apparently the signal has a 32 kHz carrier and a message consists of 17 bits (in the next chapter we will see that this is incorrect). A zero and a one bit both start with a pulse, followed by a pause. For the "one" bit the pause is longer than for the "zero" bit. Unfortunately I do not have the original remote control anymore but a clone, so we will start with trying to collect data. We would need to do that anyway because we have no information regarding the content of the messages. So we will press a few buttons and see what data we can collect.

Hardware

Before we can collect anything we need to build the hardware. The infrared signal consists of a carrier wave which is switched on and off in a specific pattern. Although it is possible to collect and decipher this signal using a program and a few components it is much easier to use a standard component for that. The main differentiation between the different types is the frequency of the carrier wave.

Available types for different carrier frequencies

Type	fo	Type	fo
TSOP1730	30 kHz	TSOP1733	33 kHz
TSOP1736	36 kHz	TSOP1737	36.7 kHz
TSOP1738	38 kHz	TSOP1740	40 kHz
TSOP1756	56 kHz		

Figure 131. The decoder type depends on the carrier wave frequency.

According to the data we found on the internet the frequency is 32 kHz, so type TSOP1733 is closest. Fortunately there appears to be some margin of error here because I happened to have a 1737 laying about and that works well too.

Figure 132. Pin layout of the TSOP 1737 IR decoder.

The decoder output must be connected to the +5 volts using a pull-up resistor, because the decoder can make the line low but not high. The signal that the decoder delivers is reversed, that means an infrared "one" will be seen as a zero. We need to remember this when we analyze the data. As a microcontroller we chose the 18f4455.

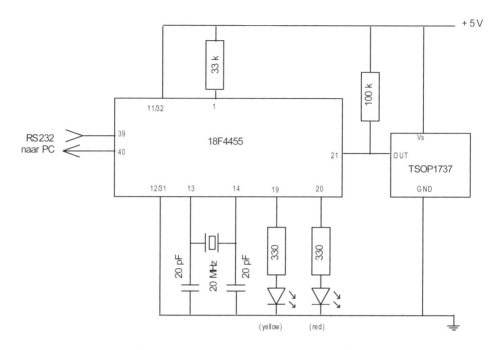

Figure 133. Schematic of the IR receiver.

Figure 134. The IR receiver with remote control.

On the photo the remote control that I use is next to the breadboard. It can be used to switch the left LED on and off.

Software

Now that the hardware is finished let's put the following program in it:

```
-- JAL 2.4i
include 18F4455_bert

-- pins
pin_d0_direction = output
pin_d1_direction = output
pin_d2_direction = input

pin_d0 = 0
pin_d1 = 0

-- variables
var byte counter[30], t

forever loop

   -- signal ready to test
   pin_d1 = 1
   for 30 using t loop
      counter[t] = 0
   end loop

   -- wait for the pin to go low
   while pin_d2 loop end loop

   t = 0
   for 15 loop

   -- measure low time
   while !pin_d2 loop
      counter[t] = counter[t]+1
      delay_10us(1)
   end loop

   -- and subsequent high time
```

```
   while pin_d2 loop
      counter[t+1] = counter[t+1]+1
      delay_10us(1)
   end loop

   t = t+2

end loop

-- signal busy
pin_d1 = 0

-- dataset complete send to PC
for 30 using t loop
   serial_sw_write(counter[t])
   delay_100ms(1)
end loop

-- make sure all data have passed
delay_1s(5)

end loop
```

The infrared decoder is connected to pin d2. Once the data array is cleared the program will wait for this pin to be low. There is a pull-up resistor to the +5 volts, so the signal is normally high.

```
while pin_d2 loop end loop
```

As soon as the pin is low we will measure 15 times the high and low times.

```
-- measure low time
while !pin_d2 loop
   counter[t] = counter[t]+1
   delay_10us(1)
end loop

-- and subsequent high time
while pin_d2 loop
   counter[t+1] = counter[t+1]+1
   delay_10us(1)
end loop
```

These times are stored in the "counter" array, and then send to the PC. There you can visualize them using MICterm or store them in a file for analyses. This is part of the data that was received after pressing the buttons for channels 1, 2, 3 and 4.

```
Button channel 1
35 168 35 67 35 67 35 67 35 67 35 168 34  67 35 67  35 67 35 67 38 64
Button channel 2
36 167 35 67 35 67 35 67 35 67 35  67 35 167 35 67  35 67 35 67 38 64
Button channel 3
36 167 35 67 38 64 35 67 35 67 35 168 35 167 35 67  38 64 35 67 38 64
Button channel 4
36 167 35 67 35 67 35 67 38 64 35  67 35  67 35 167 38 65 37 65 34 67
```

In the next Figure these data are shown as a graph. Remember that the signal is upside down because the TSOP inverts the signal.

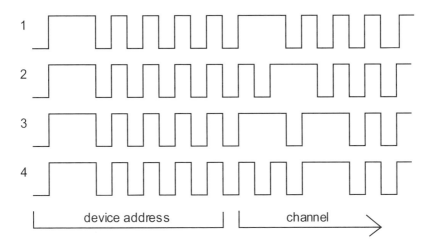

Figure 135. Received signals as a graph.

It is immediately apparent that the first five bits are always the same, apparently this is the address of my television. The next parts are different, and you can clearly recognize a binary counting pattern, but backwards.

channel	binary code
1	1000
2	0100
3	1100
4	0010

Table 31. Channel codes.

With this information we can write our program. We use the exact same routine that we used to gather a series of data. We start the array at 11 and skip the first five bits.

```
for 8 loop
    -- make room
    data = data>>1
    -- is it a 0 or a 1 (need to use an intermediary
    -- variable due to a bug in JAL)
    set = counter[t+11]
```

If the pause is longer than 100 then it is the pause that belongs to the "one" bit, so variable "incoming" is made 1.

```
    if set > 100 then
        incoming = 1
    else
        incoming = 0
    end if
    delay_10ms(1)
    t = t + 2
end loop
```

It looks as if variable incoming is not used for anything. It is defined as:

```
var bit incoming at data:7
```

That means that incoming is the leftmost bit of variable "data". So when incoming is set to one, the leftmost bit of data is also set to one. And with each loop data is shifted one position to the right

```
data = data >> 1
```

which means that after those loops all bits in variable data have received the correct value. This value is sent to the PC via the Wisp passthrough function so you can observe incoming data if you want, but that is not a requirement. If the incoming data refers to channel 1 the left LED is switched on. If it is any other channel the left LED is switched off again. Of course you can expand this program with more channels.

This part of the program may have surprised you:

```
set = counter[t + 11]
if set > 100 then
```

We could have used a single command as well:

> if counter[t + 11] > 100 then

Well, we should, but due to a bug in this version of JAL we cannot compare with data in an array, so we need to use an intermediary variable.

The right LED is a status LED that indicates if the program is analyzing incoming data. If the LED rapidly flashes data is being received.

If you press a button on the remote control too long you will start receiving odd channel numbers. This is because the remote control will repeat the message automatically, but in a different format. This way the television can tell the difference between a button that is pressed repeatedly and a button that is held down for an extended period of time. This program doesn't correct for that. The best solution is to use a single channel to switch the LED off (for example channel 2) instead of just any channel, as it is at this moment.

```
-- JAL 2.4j
include 18F4455_bert

-- pins
pin_d0_direction = output
pin_d1_direction = output
pin_d2_direction = input

pin_d0 = 0
pin_d1 = 0

-- variables
var byte counter[30], t, data, set
var bit incoming at data:7

forever loop

   -- signal ready to test
   pin_d1 = 1
   for 30 using t loop
      counter[t] = 0
   end loop

   -- wait for the pin to go low
   while pin_d2 loop end loop
```

```
-- set counter
t = 0

-- get the data from the ir sensor
for 15 loop

  -- measure low time
  while !pin_d2 loop
    counter[t] = counter[t] + 1
    delay_10us(1)
  end loop

  -- and subsequent high time
  while pin_d2 loop
    counter[t + 1] = counter[t + 1] + 1
    delay_10us(1)
  end loop

  -- increment counter
  t = t + 2

end loop

-- signal busy
pin_d1 = 0

-- extract the message and turn into a byte
data = 0
t = 0
for 8 loop
  -- make room
  data = data >> 1
  -- is it a 0 or a 1 (need to use an intermediary
  -- variable dus to a bug in JAL)
  set = counter[t + 11]
  if set > 100 then
    incoming = 1
  else
    incoming = 0
  end if
  delay_10ms(1)
```

```
      t = t + 2
   end loop

   -- set the yellow LED
   if data == 1 then
      pin_d0 = 1
   else
      pin_d0 = 0
   end if

   -- send data to PC
   serial_sw_write(data)
   delay_100ms(1)

end loop
```

9.4 IR - Transmitter (remote control)

In this project we will build an infrared transmitter that can be used to remotely control the previous project. If you own a Sharp TV you can use this project as a remote control for it.

Technical background

This time we will use an infrared LED so we will have to make our own carrier wave. The frequency should be 32 kHz for the TV but the TSOP1737 receiver expects a frequency of 37 kHz. We chose the TV frequency, because we know from the previous project that the TSOP1737 will work with that frequency as well. A smart way to make a frequency signal is with the PWM module, because once we set it we do not have to concern ourselves with it anymore. According to the datasheet of the 18f4455 the formula for setting the PWM period is (pr2+1) * 4 * Tosc * TMR2prescaler.

A frequency of 32 kHz means a period of $1/32,000 = 3.125 \cdot 10^{-5}$. So:

$$(pr2+1) * 4 * Tosc * TMR2prescaler = 3.125 \cdot 10^{-5}$$

where Tosc is the clock speed. Normally the clock speed is equal to the speed of the crystal - in our case 20 MHz - but the 18f4455 has the possibility of running faster than its clock, and this option is enabled.

That means that Tosc = 1/48,000,000, so:

$$(pr2+1) * TMR2prescaler = 3.125 \cdot 10^{-5} / (4 * 1 / 48,000,000)$$
$$= 3.125 \cdot 10^{-5} / 8.33 \cdot 10^{-8} = 375$$

Which means that:

$$(pr2+1) \, TMR2prescaler = 375$$

This is a single equation with two unknowns that cannot be solved mathematically. We simply try some possibilities to find a suitable combination. Fortunately for us the options with respect to the TMR2prescaler are limited to 1, 4, or 16. That means we have the following possible combinations:

TMR2prescaler	pr2+1
1	too large
4	93.79
16	23.44

Table 32. TMR2prescaler and pr2+1 combinations.

Of course pr2+1 needs to be a whole number, and rounding off will give a smaller error for 93.79 than for 23.44 so we select 4 for the TMR2prescaler and 94 for pr+1. That leaves us with a PWM frequency of:

$$\frac{1}{\frac{(93+1) * 4 * 4}{48000000}} = 31.9 \text{ kHz}$$

A very satisfying result, considering that our target is 32 kHz.

T2CON: TIMER2 CONTROL REGISTER

U-0	R/W-0	R/W-0	R/W-0	R/W-0	R/W-0	R/W-0	R/W-0
—	T2OUTPS3	T2OUTPS2	T2OUTPS1	T2OUTPS0	TMR2ON	T2CKPS1	T2CKPS0
bit 7							bit 0

Legend:
R = Readable bit   W = Writable bit   U = Unimplemented bit, read as '0'
-n = Value at POR   '1' = Bit is set   '0' = Bit is cleared   x = Bit is unknown

bit 7   Unimplemented: Read as '0'
bit 6-3   T2OUTPS3:T2OUTPS0: Timer2 Output Postscale Select bits
            0000 = 1:1 Postscale
            0001 = 1:2 Postscale
            •
            •
            •
            1111 = 1:16 Postscale
bit 2   TMR2ON: Timer2 On bit
            1 = Timer2 is on
            0 = Timer2 is off
bit 1-0   T2CKPS1:T2CKPS0: Timer2 Clock Prescale Select bits
            00 = Prescaler is 1
            01 = Prescaler is 4
            1x = Prescaler is 16

Figure 136. Register t2con of the 18f4455 microcontroller

The TMR2prescaler is in the t2con register, and of course we need to switch this timer on as well:

t2con = 0b_0000_0101

CCP1CON: ECCP CONTROL REGISTER (40/44-PIN DEVICES)

R/W-0	R/W-0	R/W-0	R/W-0	R/W-0	R/W-0	R/W-0	R/W-0
P1M1	P1M0	DC1B1	DC1B0	CCP1M3	CCP1M2	CCP1M1	CCP1M0
bit 7							bit 0

Legend:
R = Readable bit   W = Writable bit   U = Unimplemented bit, read as '0'
-n = Value at POR   '1' = Bit is set   '0' = Bit is cleared   x = Bit is unknown

bit 7-6 **P1M1:P1M0**: Enhanced PWM Output Configuration bits
If CCP1M3:CCP1M2 = 00, 01, 10:
xx = P1A assigned as Capture/Compare input/output; P1B, P1C, P1D assigned as port pins
If CCP1M3:CCP1M2 = 11:
00 = Single output: P1A modulated; P1B, P1C, P1D assigned as port pins
01 = Full-bridge output forward: P1D modulated; P1A active; P1B, P1C inactive
10 = Half-bridge output: P1A, P1B modulated with dead-band control; P1C, P1D assigned as port pins
11 = Full-bridge output reverse: P1B modulated; P1C active; P1A, P1D inactive

bit 5-4 **DC1B1:DC1B0**: PWM Duty Cycle Bit 1 and Bit 0
<u>Capture mode</u>:
Unused.
<u>Compare mode</u>:
Unused.
<u>PWM mode</u>:
These bits are the two LSbs of the 10-bit PWM duty cycle. The eight MSbs of the duty cycle are found in CCPR1L.

bit 3-0 **CCP1M3:CCP1M0**: Enhanced CCP Mode Select bits
0000 = Capture/Compare/PWM off (resets ECCP module)
0001 = Reserved
0010 = Compare mode, toggle output on match
0011 = Capture mode
0100 = Capture mode, every falling edge
0101 = Capture mode, every rising edge
0110 = Capture mode, every 4th rising edge
0111 = Capture mode, every 16th rising edge
1000 = Compare mode, initialize CCP1 pin low, set output on compare match (set CCP1IF)
1001 = Compare mode, initialize CCP1 pin high, clear output on compare match (set CCP1IF)
1010 = Compare mode, generate software interrupt only, CCP1 pin reverts to I/O state
1011 = Compare mode, trigger special event (CCP1 resets TMR1 or TMR3, sets CCP1IF bit)
1100 = PWM mode: P1A, P1C active-high; P1B, P1D active-high
1101 = PWM mode: P1A, P1C active-high; P1B, P1D active-low
1110 = PWM mode: P1A, P1C active-low; P1B, P1D active-high
1111 = PWM mode: P1A, P1C active-low; P1B, P1D active-low

Figure 137. The ccp1con register of the 18f4455.

The PWM must be switched on too of course, and we will do that using the ccp1con register.

    on   ccp1con = 0b_0000_1100
    off   ccp1con = 0b_0000_0000

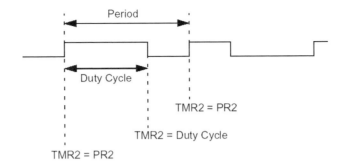

PWM Duty Cycle = CCP • TOSC • (TMR2 Prescale Value)

Figure 138. Calculation of the duty cycle.

At a frequency of 32 kHz the period is $1/32{,}000 = 3.125 \cdot 10^{-5}$. We opt for a 50% duty cycle. This means 50% of the period, so 50% of $3.125 \cdot 10^{-5}$. Entering these data in the formula from the datasheet results in 188 for ccp:

$$ccp = \frac{0.5 * 3.125 \cdot 10^{-5}}{1/48{,}000{,}000 * 4} = \frac{0.5 * 3.125 \cdot 10^{-5}}{8.33 \cdot 10^{-8}} = 188$$

In binary that is 0b_1011_1100 of which the lowest two bits must be stored in ccp1con and the remaining 8 in ccpr1l (the duty cycle is a 10 bit affair). The lowest two bits are zero, so we only need to store 0b_0010_1111 in ccpr1l.

Hardware

We will use the TSUS5202 infrared LED, with the following data according to the datasheet:

Current	150 mA (max peak 300 mA)
Forward voltage	1.3-1.7 volts
Max dissipation	170 mW

The current can be 150 mA but that is a bit much because the power regulator in the Wisp648 is not equipped with a heat sink. We will use a lower current of 75 mA. The microcontroller cannot source this because it is over the maximum of 25 mA, so we will have to use a transistor.

The voltage drop over the LED is at least 1.3 volts and for the transistor 0.7 would be a reasonable assumption. That means that the voltage over the current limiting resistor is 5 - 1.3 - 0.7 = 3 volts. Using Ohm's law the value of the resistor can be calculated. The result is 40 ohm. This value doesn't exist so we select the nearest larger one: 47 ohm. That means the current is 64 mA.

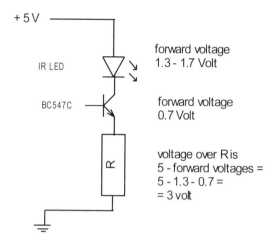

Figure 139. Calculation of the current limiting resistor of the infrared LED.

The power consumption of the LED when using this resistor is $P = V * I^2 = 5 * 0.064^2 = 0.020 = 20$ mW, well below the maximum dissipation, so we can use this resistor.

We will use the four step plan in the appendix (13.6) to select a transistor.

1a. The transistor must have an $I_{c(max)}$ of at least 64 mA.

1b. The resistance of the LED including the resistor is calculated using $R = V/I = 5/78$ mA = 64 ohm, so this is R_L

2. The maximum current the pin can deliver is 25 mA, so the transistor needs an amplification of at least $h_{FE(min)} = 5 * (64$ mA $/ 25$ mA$) = 12.8$

3. Based on this we select the BC547C transistor from the table in the appendix. It has an $I_{C(max)}$ of 500 mA and a $h_{FE(min)}$ of 420.

4. Because the supply voltage of the transistor part is identical to the rest of the project we can use the simplified formula for the base resistor.

$R_B = 0.2$ x R_L x h_{FE} so $R_B = 0.2$ x 64 x $420 = 5376$. The closest value that exists is 5k6 (green - blue - red).

At an amplification of 420 the base current of this transistor is 64/420 = 0.15 mA, way below the maximum value of 25 mA for the microcontroller.

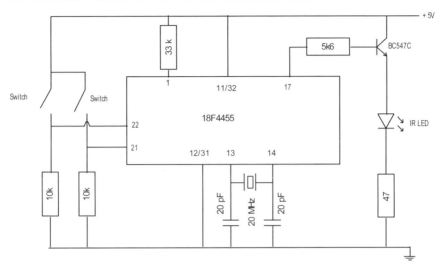

Figure 140. The schematic of the transmitter.

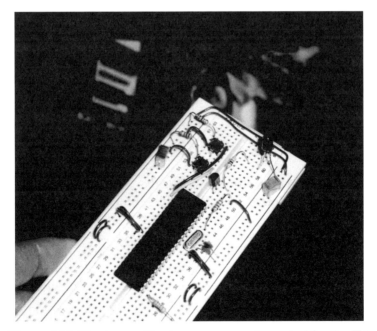

Figure 141. The transmitter in use as remote control for the TV.

Tip: you cannot see infrared light so you cannot see if the LED is actually on. Your digital camera however can. So you can look through your camera at the LED to see if it is working. Of course that trick also works with your normal remote control

Software

In the previous project we just deciphered the first part of the signal because that was enough to control our LEDs. This time however we need to know the complete signal otherwise the TV will not respond.

By collecting 60 instead of 15 bits we get the following result. You must press the "channel 1" button twice, so somewhere in this data the signal starts for the second time:

35 168 35 67 35 67 35 67 35 67 35 168 34 68 34 67 38 65 34 67 35 67 35 67 35 67 35 168 35 67 34 155 35 167 35 67 35 67 35 67 35 67 35 168 34 168 35 168 34 168 35 168 35 167 35 168 35 67 35 167 35 151 35 168 34 68 34 68 34 67 35 67 35 168 35 67 35 67 35 67 35 67 35 67 34 68 34 68 34 168 35 67 35 154 35 168 34 68 34 68 34 67 35 67 35 67 35 168 35 167 35 168 35 167 35 168 35 167

We will not make a graph this time but convert the time measurements to zeros and ones, with this result:

    10000100000001011000001111111011
    10000100000001011000001111111

The first row of 32 bits is apparently the complete message for "channel 1". If we divide the first row in two parts and put the underneath each other a pattern emerges.

    10000 10000000 101
    10000 01111111 011

The address is identical (the first five bits) but the command bits (the next 8) are all flipped. The first part is terminated with 101 and the second part with 011.

Using our newfound knowledge we should be able to predict the results for the channel two button. The address of the TV, followed by the binary number for two, but backwards, the terminating bits, and then the second row as previously discussed.

    10000 01000000 101
    10000 10111111 011

Using these two strings we should be able to write a program that can control channel 1 and 2. We will call these stings "message" and define them as a word. A word has 16 bits, and that is quite convenient because our string is exactly 16 bits per part. So for channel 1 that results in:

>   message1 = 0b_1000_0100_0000_0101
>   message2 = 0b_1000_0011_1111_1011

Sending this message is done by enabling the PWM module for 280 uS. During that time a frequency of 32 kHz is transmitted (the carrier).

>   ccp1con = 0b_0000_1100
>   delay_10us(28)
>   ccp1con = 0b_0000_0000

The next step is to check what bit we need to transmit. For this we use the variable "outgoing" that is defined as the leftmost bit of the message:

>   var bit outgoing at message:15

Please note that computers start counting at zero, so the 16th bit is number 15. When the bit we want to transmit is high we will pause 1900 uS. If it low we will pause 780 uS. Next we will shift message to get to the next.

>   if outgoing == 1 then
>   delay_10uS(190)
>   else
>   delay_10us(78)
>   end if
>   message = message << 1

This is the completed program. It can be used in combination with the previous project, or with a Sharp TV (assuming it uses this protocol). If you don't own a Sharp TV the extensive description of these projects may be useful in designing your own remote control.

>   -- JAL 2.4i
>   include 18F4455_bert
>
>   -- pins
>   pin_c2_direction = output
>   pin_d2_direction = input

```
pin_d3_direction = input

-- set PWM frequency
pr2 = 93
t2con = 0b_0000_0101
ccpr1l = 0b_0010_1111

-- declare variables
var word message,message1, message2
var bit outgoing at message:15
var byte flag,t

forever loop

  flag = 0
  -- left button
  if pin_d3 then
    -- channel 1
    message1 = 0b_1000_0100_0000_0101
    message2 = 0b_1000_0011_1111_1011
    flag = 1
  end if

  -- right button
  if pin_d2 then
    -- channel 2
    message1 = 0b_1000_0010_0000_0101
    message2 = 0b_1000_0101_1111_1011
    flag = 1
  end if

  -- something to send
  if flag == 1 then

    -- normal message
    message = message1
    for 16 loop
      -- pulse on c2
      ccp1con = 0b_0000_1100
      delay_10us(28)
      ccp1con = 0b_0000_0000
      -- pause
```

```
     if outgoing == 1 then
       delay_10uS(190)
     else
       delay_10us(78)
     end if
     message = message << 1
   end loop

   -- delay
   delay_1ms(40)

   -- inverted message
   message = message2
   for 16 loop
     -- pulse on c2
     ccp1con = 0b_0000_1100
     delay_10us(28)
     ccp1con = 0b_0000_0000
     -- pause
     if outgoing == 1 then
       delay_10uS(190)
     else
       delay_10us(78)
     end if
     message = message << 1
   end loop

   -- clear flag
   flag = 0

 end if

end loop
```

9.5 USB - Serial echo

Modern portables do not have a serial connection any more. That is a pity because a serial connection is very easy to set up and it is a very convenient way to communicate with a microcontroller. Microchip - the manufacturer of the PIC microcontrollers - has issued a USB pack with libraries and examples that can be used to make USB applications relatively easily. In this chapter we will use the JAL translations of the Microchip pack

Technical background

USB is an acronym for Universal Serial Bus. Technically speaking it is not a bus because it only allows one on one communication between the host - in our case the PC - and a device - in our case the microcontroller. If you want to connect multiple devices to the same host you need a hub. Because devices can optionally receive power through the USB bus some hubs have a built in power supply, to prevent overloading of the host. Using a hub will protect your PC so it is advisable to use one when you are planning to experiment. A device doesn't know that it is connected to a hub so both hardware and software do not have to be changed depending on whether you use a hub or not. In fact the device will see the hub as host.

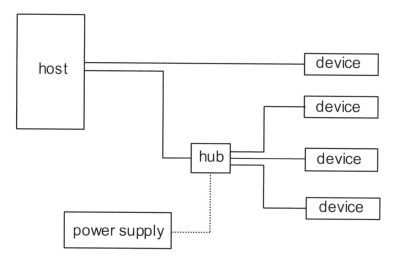

Figure 142. USB connections.

Every device that is connected using USB has to indicate what type of device it is. In this book we will use class 2 (in this project) and class 3 (in the next two projects).

Base Class	Descriptor Usage	Description
00h	Device	Use class information in the Interface Descriptors
01h	Interface	Audio
02h	Both	Communications and CDC Control
03h	Interface	HID (Human Interface Device)
05h	Interface	Physical
06h	Interface	Image
07h	Interface	Printer
08h	Interface	Mass Storage
09h	Device	Hub
0Ah	Interface	CDC-Data
0Bh	Interface	Smart Card
0Dh	Interface	Content Security
0Eh	Interface	Video
0Fh	Interface	Personal Healthcare
DCh	Both	Diagnostic Device
E0h	Interface	Wireless Controller
EFh	Both	Miscellaneous
FEh	Interface	Application Specific
FFh	Both	Vendor Specific

Table 33. Official USB classes.

A second identification is the VID/PID number; the vendor and product identification number. With this number the host can search for a driver, if appropriate. Of course this only works when every device uses a unique number. In this book we use VID and PID numbers that are owned by Microchip. You can use these numbers for free for personal use, but not in applications or devices that you want to sell. In that case you need to purchase your own VID number. In this book we use the following combinations, in hexadecimal:

project	VID	PID
Serial echo	04D8	000A
Teasing mouse	04D8	0000
A/D in Excel	04D8	0055

Table 34. Microchip VID and PID numbers.

In the section 13.3 of the appendix you will find an overview of all USB libraries and commands. In this project we will use these commands:

command	description
Serial_USB_Poll	Take care of USB business and make sure the connection is still open.
Enable_Modem	You can use this variable to check if the USB-serial link is still up.
Serial_USB_Read(Data)	Receive serial data and store it in the variable "data". You can use this command also to check if any data has arrived in the buffer.
Serial_USB_Write(Data)	Send the content of variable data.
Serial_USB_Locate(horizontal, vertical)	Move the cursor (on a VT52 terminal or emulation) to the coordinates horizontal, vertical.
Serial_USB_Clear	Clear the screen from the current cursor position (on a VT52 terminal or emulation).
Serial_USB_Home	Move the cursor to the home position - the upper left corner (on a VT52 terminal or emulation).
Serial_USB_Byte(data)	Send a byte named data as three digits - or less- instead of a single number.
Serial_USB_Printf(array)	Send a complete array - which needs to be defined first - with a single command. For example const byte mystr[] = "Bert van Dam" followed by serial_usb_printf(mystr).
Serial_USB_Disable	Disable the USB connection. Do remember to - manually - disable the connection on the PC side as well.

command	description
Serial_USB_Enable	Enable the USB connection if you have disabled it with the previous command. Make sure to wait at least 3 seconds between these two commands. Do remember to - manually - enable the connection on the PC side as well.

Table 35. Serial USB commands.

The program in this project will setup a serial USB connection. Characters typed on the PC keyboard are sent to the microcontroller and echoed back with an accompanying text, and shown on the PC screen.

It is important that the command

    serial_usb_poll

is called on a regular basis. This command takes care of USB traffic between PC and microcontroller. This traffic runs continuously, even if your program is "doing nothing". The reason for this is that a USB connection has multiple levels. As soon as the PC sees a USB device it will enable a connection for information exchange which is for example used to determine what type of device it is. This connection is always open, and there is always communication.

On top of that connection is the connection used for serial communication. This is only used when the microcontroller or PC have an actual message for each other. Because the lower level communication needs to be updated on a regular basis you cannot use long delays. You should replace those by a waiting loop that regularly issues the poll command.

The variable "enable_modem" is used to remember if the serial communication is still up and running. The best thing to do is to only send or request data after verifying that the connection is active.

```
if enable_modem then
   -- your commands
end if
```

In order for this project to work you need to install a driver on your PC. The driver is part of the download packet. You will find the instructions for the driver installation in the software section of this project.

Hardware

In order to connect the 18f4455 to the PC you need a USB plug. A convenient way is to cut a USB cable and solder pins to the wires that can be inserted in your breadboard. Make sure to cut the cable near the plug that you do not want. The connections are as follows:

number	color	purpose	pin on the 18f4455
1	red	+ power (V_{DD})	do not use
2	white	data- (D-)	23 (c4)
3	green	data+ (D+)	24 (c5)
4	black	ground	0 on the breadboard

Table 36. Connections of the USB plug.

If your cable has different colors then use a multimeter and the following Figure to determine which wire has to go where.

4　3　2　1
GND　D+　D-　Vdd

Figure 143. USB plug, white plastic at the bottom.

You will notice by the way that the ground and power pins slightly protrude. This way the device will be connected to the power before any of the data pins are connected. That means that USB devices are hot swappable, meaning that the device can be removed while the power is on, in fact even when the communication is running[34]. That also explains the need for regular polling: the user may have unplugged the device in the mean time.

USB uses 3.3 volts while the microcontroller runs on 5 volts. The 18f4455 has its own USB power supply on board. For that reason we will not use the V_{DD} connection in the USB plug. The stabilizer that the 18f4455 uses for this needs a capacitor, but that is impossible inside a chip. For that reason a capacitor needs to be connected to pin 18 - V_{USB}. This explains why pin c3 doesn't exist: this pin is part of the USB power supply.

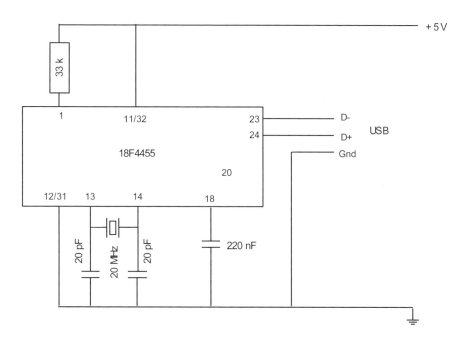

Figure 144. USB echo schematic.

The thick wire on the lower right is the USB cable going to the PC (or better yet: a powered hub). Do not plug the project in just yet, first read the software section which contains driver installation instructions.

---

[34] This is of course from a hardware point of view only. Passing data may get corrupted which may cause all sorts of nasty problems.

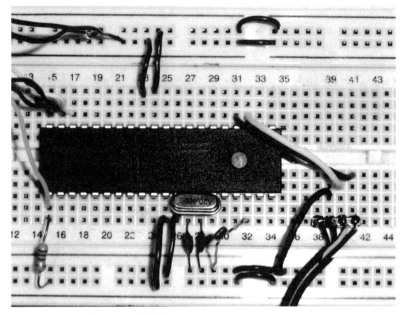

Figure 145. The project on a breadboard.

Software

Do not plug the project in just yet. Put this program into the microcontroller:

```
--JAL2.4j
include 18f4455_bert
include usb_rs232

-- variables
var byte data
const byte mystr[] = "You typed key: "

-- main loop
forever loop

   -- serve USB requests and confirm connection is still open
   serial_usb_poll

   if enable_modem then
      -- the rs232 connection is up and running
      if serial_usb_read(data) then
```

```
          -- if data is available, then echo
          serial_usb_locate(3,5)
          serial_usb_printf(mystr)
          serial_usb_write(data)
       end if
    end if

    end loop
```

This program needs a driver. You can only install this driver if you are the administrator on your PC, or if you have administrator rights. If this is your own PC and you have no clue what this is about, but you can normally install software yourself you already have these rights. If not you need to log in as administrator before continuing.

Because a USB driver needs to be installed per USB connection it may be convenient to remember with USB port you use for this project. The next time you can use the same port so you do not have to install drivers all the time.

Insert the USB cable into the PC and put power on the breadboard. After a short while a message will pop up on your PC saying that new hardware has been found, with description "CDC RS-232 emulation demo". When your computer, or rather Windows, discovers that this is a class 2 device - serial communication - it will search for the correct driver using the VID/PID numbers. Since this is the first time you use this project the search will fail, and Windows will suggest that a driver should be installed. Do not give Windows permission to look for this driver on the internet, and point it to the directory where the driver is located (c:\picdev2\project\gereed\9.5\driver). You will receive a rather threatening warning that the driver isn't certified. That is correct so ignore the warning and select "continue anyway".

There is a small chance that you are asked for the file mm_mchpusb.sys. This file is in the driver directory. On a Windows XP system that is the directory c:\windows\system32\drivers. Otherwise use the search function to locate this driver.

Now wait for the installation process to complete. Windows will report that it is done with a window that contains the name of the new driver.

On your PC you now have a new COM port. Not a real one, but a virtual one, but to the programs on your PC it looks just like the real thing. Which number has been assigned to this port is different on every PC, for it depends on the number of ports your PC has, and on the number of virtual drivers that you have loaded at some point. Start the device manager and look for the section "ports, COM & LPT". On a Windows XP machine the device manager can be accessed using start - this computer - system information - hardware - device manager.

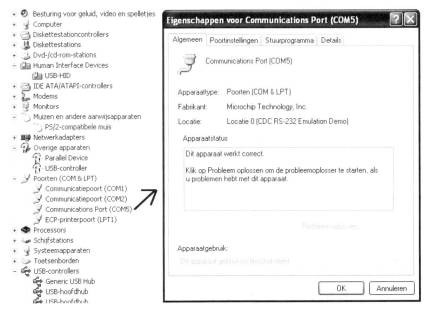

Figure 146. Win XP device manager (Dutch version, the new port is COM5).

There you will find a new COM port. Click on this entry to verify that it is the correct one. At "location" it should say (amongst other things) "CDC RS-232 Emulation Demo". As you can see in the previous Figure on my PC this is port COM5.

Please note: this port only exists when the driver is loaded, and the driver will only be loaded when your project is connected and powered.

Open a communications program that can handle VT52 terminal emulation, for example HyperTerm. Select the correct COM port and use the following settings:

      Databits   8
      Parity   none
      Stopbits   1
      Datatransportcontrol   hardware

You can use any speed you like, the USB connection will adapt automatically!

Figure 147. The result of our program.

Now press a key. On the PC the text "You typed key: " will appear followed by the key you just pressed.

Note that the Windows dong-ding sound only means that Windows has seen the device, not that it is actually operational. After the dong-ding wait 8 seconds before you use the port. If you let HyperTerm connect too quickly you will get the message that the port doesn't exist. In that case wait a bit and then try again.

9.6 USB - Teasing mouse

In this project we will make an annoying program to fool PC users, by making their mouse pointer jump to the right at random moments. You do not need to install any software on the PC, just insert the USB cable and the teasing can start.

Technical background

In the previous project a driver was required. The Windows operating system is equipped with a series of default drivers for certain devices. For these devices you do not need to install a driver. You simply plug it in, and the device will be recognized all by itself. This applies for example to a mouse - in this project - or a keyboard - as used in the next project. Devices that belong to this class are called HID, Human Interface Device (class 3). In this project we will use the simplified version, the so called boot version because all Windows versions support this version by default.[35]

---

[35] Boot refers to booting - or starting up - of the system.

The communication between a HID device and the host is carried out using reports. A report is a series of data that are sent as a group in a predetermined format.

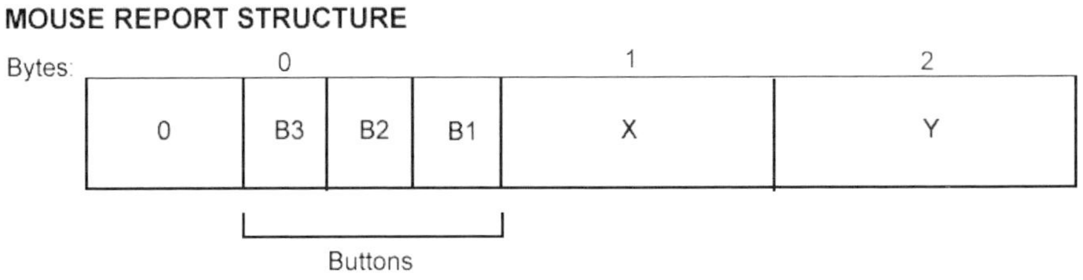

MOUSE REPORT STRUCTURE

Bytes:	0				1	2
	0	B3	B2	B1	X	Y

(B3, B2, B1 = Buttons)

Figure 148. Mouse report from Microchip application note AN1163[36].

The previous Figure shows the mouse report that a mouse that is connected to a USB port is expected to send to the host. The report consists of three bytes. In the first byte (byte 0) the first three bits indicate the position of three mouse buttons.

    B1 = the left mouse button
    B2 = the right mouse button
    B3 = the mouse button "hidden" in the little wheel in the center

The next two bytes indicate how much the mouse has been moved in the x direction - horizontal on the screen - and the y direction - vertical on the screen - since the previous report.

In the software the mouse report is defined as a three byte array:

    data_buffer [0]
    data_buffer[1]
    data_buffer[2]

Let's say we want to tell the PC that the center button - the one under the mouse wheel - has been pressed. We will send the following report:

    data_buffer[0] = 0b_0000_0100
    data_buffer[1] = 0
    data_buffer[2] = 0
    hid_mouse_write

---

[36] AN1163: USB HID Class on an Embedded Device, from www.microchip.com

The "hid_mouse_write" command takes care of the actual transmission of the report. Instead of 0b_0000_0100 you can of course also write 4. To indicate that the button has been released you need to send an empty report. Between the two reports you should give the PC time to process the first report. In practice about 20 mS is enough. You must use a special delay version to make sure that the connection is kept open.

> usb_delay_1ms(20)

If for example you want to momentarily press the right mouse button to open the context sensitive help menu on the PC you can use the following commands:[37]

> -- click the mouse button
> data_buffer[0] = 2
> data_buffer[1] = 0
> data_buffer[2] = 0
>
> -- send the report
> hid_mouse_write
> usb_delay_1ms(20)
>
> -- release the mouse button
> data_buffer[0] = 0
> data_buffer[1] = 0
> data_buffer[2] = 0
>
> -- send the report
> hid_mouse_write

Hardware

The hardware is identical to the previous project. If you haven't made a USB plug yet you can find instructions there too.

---

[37] In most Windows compliant PC programs the right mouse buttons opens a context sensitive help menu. Of course in some programs this button may do something completely different, or maybe even nothing at all.

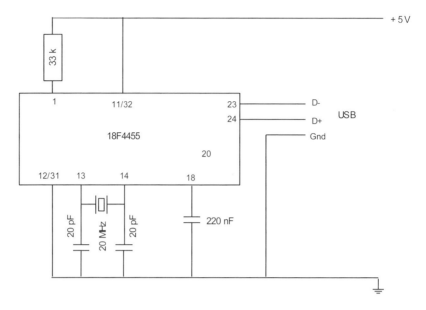

Figure 149. USB echo schematic.

The thick wire on the lower right is the USB cable going to the PC, or better yet: a powered hub.

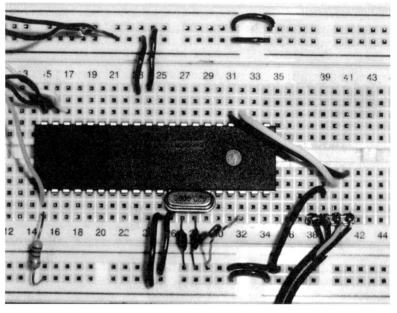

Figure 150. The project on a breadboard.

Software

The program in this book unexpectedly moves the mouse pointer to the right, at random moments. Hence the name teasing mouse. For those random moments we use the "dice" command that simulates the rolling of a dice. Of course the numbers that this command generates are not really random. We need to tell the microcontroller how to do generate these numbers using a formula, and if it has a formula it is by definition not random. So the numbers generated by dice do contain logic, but it is so complicated that the user will not recognize it. So for all practical purposes this is random enough.

Every second we throw the dice, like this:

    moveme = dice

If the result is equal to six the mouse pointer on the PC screen is moved to the right. If the result is another number the program will do nothing, and simply try again after another second.

    usb_delay_100ms(10)

Moving the mouse pointer is done by entering a movement in the mouse report. In this case 80 units to the right. You may need to experiment with this value. Some people told me that on some PC's 80 propels the pointer straight off the screen and suggested that 20 may be a better value.

    data_buffer[0] = 0
    data_buffer[1] = 80
    data_buffer[2] = 0
    hid_mouse_write

You do not need to follow this up with an empty report because the mouse movement is not an absolute value, but relative to the position in the previous report. So this doesn't translate to "move the mouse pointer to location 80,0" but simply to "move the mouse pointer 80 to the right from where ever the current position on the screen is".

The teasing mouse works independently of the normal mouse that the user uses. While the user is working on the PC the mouse pointer just suddenly moves a bit.

This is the completed program:

    --JAL2.4j
    include 18f4455_bert

```
include usb_hid_mouse

-- variables
var byte moveme

-- main loop
forever loop

    -- take care of USB business
    usb_tasks

    -- throw the dice
    moveme = dice

    -- dice is a 6
    if moveme == 6 then

        -- move the mouse
        data_buffer[0] = 0
        data_buffer[1] = 80
        data_buffer[2] = 0

        -- send the report
        hid_mouse_write

    end if

    -- 1 second delay
    usb_delay_100ms(10)

end loop
```

When you connect the hardware and switch the power on Windows will tell you that new hardware has been found. Simply wait, and after a short while you will get the message that the hardware has been installed and that you can use it. You do not need administrator privileges for this.

Note that the Windows dong-ding sound only means that Windows has seen the device, not that it is actually operational. After the dong-ding wait 8 seconds before you use USB.

In this program we make use of the following functions from the HID mouse library. A complete overview of all functions can be found in the "other libraries" section in the appendix.

command	description
HID_Mouse_Write	Send the mouse report to the PC
usb_delay_1ms(n)	Wait n (1 to 255) times 1 ms, while the USB connection is kept active.
usb_delay_100ms(n)	Wait n (1 to 255) times 100 ms, while the USB connection is kept active.
usb_tasks	Take care of USB business and make sure the connection is still open.
usb_is_configured	Check to see if the USB mouse link is still up.
usb_initialized	A flag to remember if usb_tasks needs to run the SUB setup or not.

Table 37. The HID mouse library commands.

And this command is from the random library:

command	description
dice	Select a random number from the range 1 to 6, meant for the simulation of dice.

Table 38. One of the commands from the random library.

9.7 USB - A/D measurements in Excel

In this project A/D measurements - of a potmeter - are entered straight into an Excel spreadsheet using a HID keyboard interface. The only thing you need to do yourself is turn them into a graph or perform another analyses.

Technical background

Just as in the mouse project we make use of a HID, Human Interface Device (class 3), but this time the HID keyboard. In this project we will use the simplified version, the so called boot version because all Windows versions support this version by default.[38]

The communication between a HID device and the host is carried out using reports. A report is a series of data that are sent over as a group in a predetermined format.

In this project we will use the HID keyboard send report. This report contains the data that the microcontroller will send to the PC. Because we use a standard HID interface we cannot change the format of the report The report contains eight fields. The first field contains the modifiers. These are special keys that "modify" the behavior of other keys such as Alt, Ctrl et cetera. The second field is reserved and should contain a zero. The remaining fields are used for key entries.

Offset	Field	Length (bytes)	Description
0	modifiers	1	status of special keys such as Alt, Ctrl etc
1	reserved	1	reserved (0x00)
2	keycodes[0]	1	keyboard code 1
3	keycodes[1]	1	keyboard code 2
4	keycodes[2]	1	keyboard code 3
5	keycodes[3]	1	keyboard code 4
6	keycodes[4]	1	keyboard code 5
7	keycodes[5]	1	keyboard code 6

Table 39. Keyboard send report.

In a normal keyboard two codes are connected to each key. The first code is sent when the key is pressed, the second code is sent when the key is released - the break code. This way the computer can tell the difference between a key that is pressed repeatedly and a key that is pressed continuously.[39] The next table shows some examples.

---

[38] Boot refers to booting - or starting up - of the system.
[39] If you press the key long enough it will be repeated, but not by your keyboard but by the operating system on the PC (Windows).

Key	Make code	Break code
A	1C	F0,1C
5	2E	F0,2E
F10	09	F0,09
Right arrow	E0,74	E0,F0,74
Right Ctrl	E0,14	E0,F0,14

Table 40. Some examples of key codes in hex.

Fortunately there is an easier way. If we send an empty report after a key press report the PC will regard this as a key break code regardless of the key that was pressed. So in this project we do not worry about break codes, we simple send the key press code (often called the make code) followed by an empty report.

You may have noticed that the key codes do not match the ASCII codes that you are used to. The next table shows the make codes - or keyboard scan codes as they are often called - for the keys that we need in this project. You will find a complete overview in section 13.5 of the appendix.

key	code 1	code 2	code 3
! 1	2	22	30
@ 2	3	30	31
# 3	4	38	32
$ 4	5	37	33
% 5	6	46	34
^ 6	7	54	35
& 7	8	61	36
* 8	9	62	37
(9	10	70	38
) 0	11	69	39
Enter	27	90	40

Table 41. Keyboard scan codes for this project.

There are several different keyboard codes in use. The previous table shows three often used ones. In this project we will use keyboard code column 3 which appears to be most often used. But it is possible that your keyboard uses different codes, meaning that you will need to use a different column.

An easy way to find out which codes your keyboard uses if yours are not in the table is to put this program into the microcontroller.

```
--JAL2.4j

include 18f4455_bert
include usb_hid_keyboard

-- variables
var byte counter = 0

-- pins
pin_d2_direction = input
var bit button is pin_d2
pin_d1_direction = output
var bit led is pin_d1

-- led off
led = off

forever loop

   -- take care of USB business
   usb_tasks

   -- button pressed
   if button & usb_is_configured then
     -- send to PC
     hid_keyboard_key(counter)
     serial_sw_write(counter)
     -- next number
     counter = counter + 1
     -- signal LED
     led = on
     while button loop
     end loop
   else
     led = off
   end if

end loop
```

Every time you press the button a counter is incremented and send to the PC to a terminal program - so you can see which number is sent - as well as to the keyboard buffer - so you can see which key goes with that number.

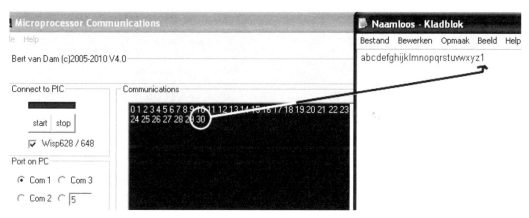

Figure 151. On my PC hid_keyboard_key(30) represents key number 1.

Take the following steps in order to use this program:

1. Build the hardware that goes with this project (if you want you can leave out the potmeter at this point), and download the previous program into it. Do not remove the Wisp programmer. Connect the USB plug to the PC.
2. Wait for the dong-ding, and then wait for another 8 seconds.
3. Disconnect the yellow Wisp wire and make sure it doesn't touch anything!
4. Start MICterm and connect at 1200 baud. Select option "RAW".
5. Open a new document in a word processor, or for example Windows Notepad, and make sure the cursor is on it (by clicking in the empty sheet).
6. Press the button on the breadboard. In MICterm the number zero appears. In the word processor something may or may not appear. Keep pressing the button until the number one appears on the word processor sheet. In MICterm you can now see which number matches key scan code 1. In the library usb_hid_keyboard search for the procedure usb_hid_byte and replace in the command "hid_keyboard_key(zw+29)" the number 29 by the difference between the number in MICterm and the number you see in the word processor. For example: if MICterm shows 34 and the number in the word processor is 1 than replace 29 by 34 - 1 = 33 so the command becomes "hid_keyboard_key(zw+33)"

Hardware

The USB connection is identical to the serial echo project. If you haven't made a USB plug yet you can find instructions there too.

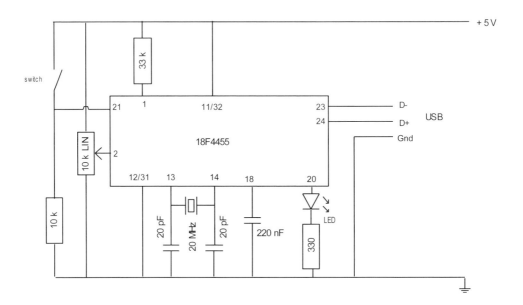

Figure 152. Schematic USB measurement in Excel.

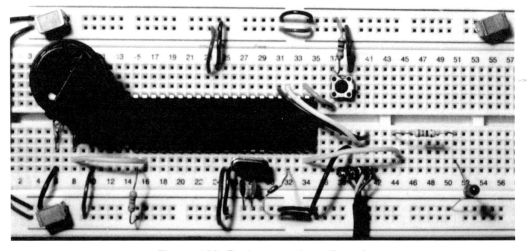

Figure 153. Project on a breadboard.

Software

When the button on the breadboard is pressed the program will start taking A/D measurements and send them to the PC. Because the USB connection is configured like a

HID keyboard these measurements look like ordinary keyboard entries. Pretty much as if a human user is entering numbers manually.

command	description
usb_is_configured()	Check to see if the USB keyboard link is still up.
usb_tasks	Take care of USB business and make sure the connection is still open.
hid_keyboard_byte(data)	Send a byte named data as three digits - or less- instead of a single number.
hid_keyboard_key(data)	Send data to the PC as byte in a keyboard report (used to send a key to the PC).

Table 42. Command from the hid-keyboard library.

Using the commands from the table the measurement can be sent to the PC, followed by the enter key: hid_keyboard_key(40). In Excel this makes the cursor jump automatically to the cell below.

```
--JAL2.4j

include 18f4455_bert
include usb_hid_keyboard

-- variables
var byte resist

-- pins
pin_d2_direction = input
var bit button is pin_d2
pin_d1_direction = output
var bit led is pin_d1

-- led off
led = off

forever loop
```

```
   -- take care of USB business
   usb_tasks()

   -- start switch pressed
   if button & usb_is_configured() then
      -- get resistance on AN0
      resist = adc_read_low_res(0)
      -- send to keyboard buffer
      hid_keyboard_byte(resist)
      -- enter to get to the next line
      hid_keyboard_key(40)
      -- signal LED
      led = on
   else
      led = off
   end if

end loop
```

Instructions

1. Download the program into the microcontroller, connect the USB cable and wait for the dong-ding. Then wait for another 8 seconds.
2. Open Excel and put the cursor in the first cell (A1). If you don't have Excel you can use another spreadsheet. You can also use a word processor or Notepad, but then you will of course not be able to process the data.
3. Press the button, and turn the potmeter. The measurements automatically appear in the leftmost column of the spreadsheet. When you release the button the measurements stop.
4. Now you can process the data, for example using a graph.

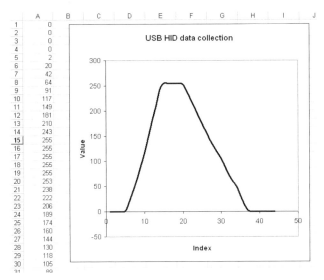

Figure 154. Result of the program in a graph.

9.8 CAN bus - Loopback

CAN bus (Controller Area Network) was designed by Bosch in 1986 to solve the problem of the ever increasing number of wires and protocols in automobiles. CAN bus is very robust and relatively insensitive to (electrical) noise which means it is often used in industry as well. CAN bus messages consist of an ID number and a maximum of 8 bytes of payload. The interesting thing about CAN bus is that the ID is the ID of the message, not of the device that the message is meant for. Any device can pick up any message if the ID shows it has some interest for it. Let's assume for example that you want to add a third brake light to an existing car. All you need to do is make sure that the CAN bus part of this brake light responds to the same message ID's as the normal brake lights.

The 18f4685 microcontroller has CAN bus functionality built in so we will use it in the next two projects.

In the first project we will built a loopback project. A single 18f4685 is communication with itself. That seems rather useless but it is an ideal way of testing CAN bus because transmitter as well as receiver are in the same microcontroller which simplifies debugging.

Technical background

CANCON: CAN CONTROL REGISTER

Mode 0	R/W-1	R/W-0	R/W-0	R/S-0	R/W-0	R/W-0	R/W-0	U-0
	REQOP2	REQOP1	REQOP0	ABAT	WIN2	WIN1	WIN0	—

Mode 1	R/W-1	R/W-0	R/W-0	R/S-0	U0	U-0	U-0	U-0
	REQOP2	REQOP1	REQOP0	ABAT	—	—	—	—

Mode 2	R/W-1	R/W-0	R/W-0	R/S-0	R-0	R-0	R-0	R-0
	REQOP2	REQOP1	REQOP0	ABAT	FP3	FP2	FP1	FP0
	bit 7							bit 0

Legend:		
R = Readable bit	S = Settable bit	
-n = Value at POR	W = Writable bit	U = Unimplemented bit, read as '0'
	'1' = Bit is set	'0' = Bit is cleared x = Bit is unknown

bit 7-5   **REQOP2:REQOP0:** Request CAN Operation Mode bits
        1xx = Request Configuration mode
        011 = Request Listen Only mode
        010 = Request Loopback mode
        001 = Request Disable mode
        000 = Request Normal mode

bit 4   **ABAT:** Abort All Pending Transmissions bit
        1 = Abort all pending transmissions (in all transmit buffers)
        0 = Transmissions proceeding as normal

bit 3-1   Mode 0:
        WIN2:WIN0: Window Address bits
        These bits select which of the CAN buffers to switch into the access bank area. This allows access to the buffer registers from any data memory bank. After a frame has caused an interrupt, the ICODE3:ICODE0 bits can be copied to the WIN3:WIN0 bits to select the correct buffer. See Example 23-2 for a code example.
        111 = Receive Buffer 0
        110 = Receive Buffer 0
        101 = Receive Buffer 1
        100 = Transmit Buffer 0
        011 = Transmit Buffer 1
        010 = Transmit Buffer 2
        001 = Receive Buffer 0
        000 = Receive Buffer 0

bit 0   **Unimplemented:** Read as '0'

bit 4-0   Mode 1:
        Unimplemented: Read as '0'
        Mode 2:
        FP3:FP0: FIFO Read Pointer bits
        These bits point to the message buffer to be read.
        0111:0000 = Message buffer to be read
        1111:1000 = Reserved

Figure 155. The cancon register of the 18f4685 microcontroller.

CAN bus is a not very complicated technique but it does require a lot of steps. This program is based on an example of Bobby Garrett and adapted to JAL and the 18f4685. We will discuss the registers that need to be set one by one. To keep it simple we will use legacy mode (mode 0).

CAN bus settings can only be modified if the CAN bus unit is in configuration mode, which is bit 7 of the cancon register. For now we will ignore the rest of the bits in this register.

    cancon = 0b_10000000

We can only do this if the unit is not busy. That is of course the case because this is the very first command, but just to make sure we will wait until the CAN bus unit confirms that has in fact entered configuration mode. The status of the CAN bus is in the canstat register.

Bits 7-5 will be 100 if the unit is in configuration mode so we will wait for that to happen:

```
-- select CAN config mode
cancon = 0b_1000_0000
while (canstat & 0b_1110_0000)!=(0b_1000_0000) loop end loop
```

The next step is to set the speed of the bus, for which three registers are used: brgcon1 t/m 3.

```
-- set speed to 100k bit
brgcon1 = 0b_0000_0011
brgcon2 = 0b_1011_1111
brgcon3 = 0b_0000_0111
```

The TX pin is default in tri-state mode (floating). That can lead to nasty problems if other wires get to close to the TX wire, so it is better to force the pin high. This can be done in the ciocon register.

CANSTAT: CAN STATUS REGISTER

Mode 0	R-1	R-0	R-0	R-0	R-0	R-0	R-0	U-0
	OPMODE2[1]	OPMODE1[1]	OPMODE0[1]	—	ICODE3	ICODE2	ICODE1	—

Mode 1,2	R-1	R-0	R-0	R-0	R-0	R-0	R-0	R-0
	OPMODE2[1]	OPMODE1[1]	OPMODE0[1]	EICODE4	EICODE3	EICODE2	EICODE1	EICODE0
	bit 7							bit 0

Legend:
R = Readable bit   W = Writable bit   U = Unimplemented bit, read as '0'
-n = Value at POR   '1' = Bit is set   '0' = Bit is cleared   x = Bit is unknown

bit 7-5   **OPMODE2:OPMODE0:** Operation Mode Status bits[1]
   111 = Reserved
   110 = Reserved
   101 = Reserved
   100 = Configuration mode
   011 = Listen Only mode
   010 = Loopback mode
   001 = Disable/Sleep mode
   000 = Normal mode

bit 4   Mode 0:
   Unimplemented: Read as '0'

bit 3-1   **ICODE3:ICODE1:** Interrupt Code bits
   When an interrupt occurs, a prioritized coded interrupt value will be present in these bits. This code indicates the source of the interrupt. By copying ICODE3:ICODE1 to WIN2:WIN0 (Mode 0) or EICODE4:EICODE0 to EWIN4:EWIN0 (Mode 1 and 2), it is possible to select the correct buffer to map into the Access Bank area. See Example 23-2 for a code example. To simplify the description, the following table lists all five bits.

	Mode 0	Mode 1	Mode 2
No interrupt	00000	00000	00000
Error interrupt	00010	00010	00010
TXB2 interrupt	00100	00100	00100
TXB1 interrupt	00110	00110	00110
TXB0 interrupt	01000	01000	01000
RXB1 interrupt	01010	10001	-----
RXB0 interrupt	01100	10000	10000
Wake-up interrupt	00010	01110	01110
RXB0 interrupt	-----	10000	10000
RXB1 interrupt	-----	10001	10000
RX/TX B0 interrupt	-----	10010	10010
RX/TX B1 interrupt	-----	10011	10011[2]
RX/TX B2 interrupt	-----	10100	10100[2]
RX/TX B3 interrupt	-----	10101	10101[2]
RX/TX B4 interrupt	-----	10110	10110[2]
RX/TX B5 interrupt	-----	10111	10111[2]

bit 0   **Unimplemented:** Read as '0'
bit 4-0   Mode 1, 2:
   EICODE4:EICODE0: Interrupt Code bits
   See ICODE3:ICODE1 above.

Figure 156. The canstat register of the 18f4685.

CIOCON: CAN I/O CONTROL REGISTER

U-0	U-0	R/W-0	R/W-0	U-0	U-0	U-0	U-0
—	—	ENDRHI[(1)]	CANCAP	—	—	—	—

bit 7   bit 0

Legend:
R = Readable bit   W = Writable bit   U = Unimplemented bit, read as '0'
-n = Value at POR   '1' = Bit is set   '0' = Bit is cleared   x = Bit is unknown

bit 7-6   **Unimplemented:** Read as '0'
bit 5   **ENDRHI:** Enable Drive High bit[(1)]
               1 = CANTX pin will drive VDD when recessive
               0 = CANTX pin will be tri-state when recessive
bit 4   **CANCAP:** CAN Message Receive Capture Enable bit
               1 = Enable CAN capture, CAN message receive signal replaces input on RC2/CCP1
               0 = Disable CAN capture, RC2/CCP1 input to CCP1 module
bit 3-0   **Unimplemented:** Read as '0'

Figure 157. The ciocon register of the 18f4685.

```
-- make sure TX doesn't get a tristate
ciocon = ciocon|0b_0010_0000
```

This concludes the setup, so now the unit can be switched to loopback mode. This time it is important to wait for the status change!

```
-- setup is done, switch to loop-back mode
cancon = 0b_01000000
while (canstat &  0b_1110_0000)!= (0b_0100_0000) loop end loop
```

Only now should we set the pin directions. If we do that sooner we will be unable to change the CAN bus mode!! This particularly seems to be the case with pin b2.

```
-- define the pins
pin_b2_direction = output
pin_b3_direction = input
pin_d2_direction = output
pin_d1 = 0
```

CAN bus uses a number of buffers. These buffers are used to send and receive messages. In this project we will use buffer 0 for sending and for receiving. First we set the transmit buffer:

```
-- transmit buffer
txb0con = 0b_0000_0000
```

TXBnDLC: TRANSMIT BUFFER n DATA LENGTH CODE REGISTERS [0 ≤ n ≤ 2]

U-0	R/W-x	U-0	U-0	R/W-x	R/W-x	R/W-x	R/W-x
—	TXRTR	—	—	DLC3	DLC2	DLC1	DLC0
bit 7							bit 0

Legend:		
R = Readable bit	W = Writable bit	U = Unimplemented bit, read as '0'
-n = Value at POR	'1' = Bit is set	'0' = Bit is cleared x = Bit is unknown

bit 7   **Unimplemented:** Read as '0'
bit 6   **TXRTR:** Transmit Remote Frame Transmission Request bit
            1 = Transmitted message will have TXRTR bit set
            0 = Transmitted message will have TXRTR bit cleared
bit 5-4   **Unimplemented:** Read as '0'
bit 3-0   **DLC3:DLC0:** Data Length Code bits
            1111 = Reserved
            1110 = Reserved
            1101 = Reserved
            1100 = Reserved
            1011 = Reserved
            1010 = Reserved
            1001 = Reserved
            1000 = Data length = 8 bytes
            0111 = Data length = 7 bytes
            0110 = Data length = 6 bytes
            0101 = Data length = 5 bytes
            0100 = Data length = 4 bytes
            0011 = Data length = 3 bytes
            0010 = Data length = 2 bytes
            0001 = Data length = 1 bytes
            0000 = Data length = 0 bytes

Figure 158. The txbndlc register of the 18f4685.

For the length of the packets we select the maximum of 8 data bytes. We do this in the transmit register of buffer 0 (txb0dlc) and the receive register of buffer 0 (rxb0dlc).

```
-- transmit and receive message length 8 bytes
txb0dlc = 0b_00001000
rxb0dlc = 0b_00001000
```

We will use the standard identifier bits:

```
-- receive valid messages with standard id in buffer 0 using filter 0
rxb0con = 0b_00100000
```

Now we are ready to define our first message. Each message needs an identification number, or ID. A receiver on the CAN bus will listen to all messages and select the messages that have an ID that is important for that unit. So it is very well possible to have multiple units on the bus all listening to the same messages. In a car for example both headlights and both taillights listen to, and act upon, messages such as "lights on". By the same token it is also possible that nobody is listening. That sounds odd but is does mean that the system is very stable. If a CAN bus unit malfunctions the rest of the bus will remain operational. In this project we can select any number we want of course.

```
-- set message identifier (id)
txb0sidl = 0b_00100000
txb0sidh = 0b_00000000
```

Next we put some data in buffer 0, for example 25.

```
-- use value 25
data = 25

-- put this into buffer 0
txb0d0 = data
```

And then we give the command to send the message:

```
-- transmit, priority level 1
txb0con = 0b_00001001
while(txb0con & 0b_0000_100) == 0b_0000_100 loop end loop
```

Receiving a message is just as simple. First we clear variable "data" to make sure that the content of that variable has really been received. Next we wait for a message to show up in buffer 0.

```
-- clear data
data = 0

-- wait for messages in buffer 0
while (rxb0con & 0b_1000_0000) != 0b_1000_0000 loop end loop
```

First we check to see if the message has the right ID number. In our case it must be identical to the ID number of the message we just send because this is of course the same message. If that is the case we copy the content out of the buffer. In this project we send the content to the PC for reference, and use it to light a yellow LED if the data is the same as the data that we sent, proving that everything works as it should.

```
-- does this message have the right ID?
if (txb0sidl == rxb0sidl) & (txb0sidh == rxb0sidh) then
  -- get the data
  data = RXB0D0

  -- send to the PC
  serial_sw_write(data)

  -- light the LED if appropriate
  if data == 25 then
    pin_d2 = high
  else
    pin_d2 = low
  end if

  -- clear the flag
  rxb0con = rxb0con & 0b_0111_1111
```

Then we clear the receipt flag to the next loop. This completes the program: we can now send and receive CAN bus messages.

At this moment all messages end up in buffer 0. Once one is in the buffer we check to see if it has the right ID. Optionally you can use a mask for each buffer to make sure that only messages that are interesting end up in the buffer. This will cut down on processing time, and it prevents that new and possibly uninteresting messages overwrite old but interesting messages that maybe haven't been processed yet. For this you can use the rxm0sidl and rxm0sidh registers for buffer 0.

Hardware

The hardware is a simple 18f4685 with a LED. You will not see anything of the CAN bus loopback because it takes place internally. So you do not need wire loops.

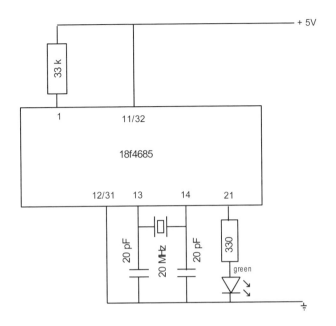

Figure 159. CAN bus loopback schematic.

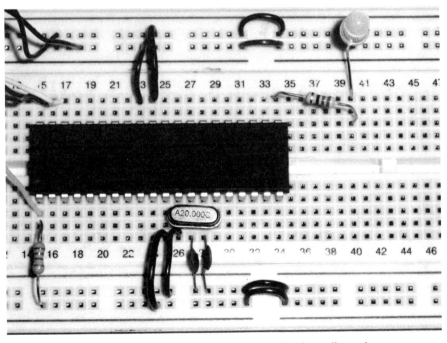

Figure 160. CAN bus loopback on the breadboard.

Software

The software has been discussed line by line in the technical background section so we will not repeat it here. What remains is the check if the right data content has been received.

```
-- send to the PC
serial_sw_write(data)

-- light the LED if appropriate
if data == 25 then
  pin_d2 = high
else
  pin_d2 = low
end if
```

The download packages contains the complete source code.

9.9 CAN bus - Remote LED

In this project we will build two CAN bus nodes. Node 1 has a button and a LED, node 2 has just a LED. The button on node 1 controls the LED on node 1 and via CAN bus also the LED on node 2.

Technical background

The CAN bus unit of the 18f4685 generates CAN bus messages, but is not capable of generating the correct voltages and currents for the actual CAN bus itself. The voltages can be up to 16 volts, and the current op to 100 mA, so we need to use CAN bus drivers - also called transceivers. In this project we will use the MCP2551 from Microchip. These driver IC's are also capable of protecting the microcontroller against voltage spikes that may occur in an automobile. A car engine is a huge source of electrical noise!

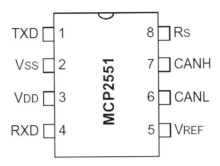

Figure 161. Pin layout of the CAN bus transceiver MCP2551.

The CAN bus itself consists of twisted pair wiring, where all CAN H pins are connected to each other, and all CAN L pins are connected to each other. You are not supposed to connect CAN L to CAN H. Both ends of the bus need a 120 ohm terminating resistor. Since we only have two nodes each node will have such a resistor. If you plain to use long(er) wires then shielded types are recommended.

The MCP2551 is connected directly to the two CAN bus pins of the 18f4455 and uses a supply voltage V_{DD} of 5 volts.

Hardware

Please note: do not built this project on a single breadboard. The MCP2551 causes nasty noise on the power supply lines. If they are too close to each other they will interfere with each other's operation. If that happens you will see that the remote LED will go on, but not off. Due to the interference the chip cannot see its own message on a hardware level and thus assumes that transmission failed. For that reason it will keep on transmitting the same "LED on" message over and over. This causes a bus overflow and eventually the bus will go down due to too many errors (over 127). The next Figure shows what this looks like on the software oscilloscope WinOscillo. The bus is filled with the same message. Please note that the bus can carry more than 5 volts, so start WinOscillo with the knob on the special hardware all the way to zero and slowly turn it up until the signal is visible. Failure to do it this way may cause too high a voltage on the microphone input on your PC, thus causing damage to your soundcard or PC.

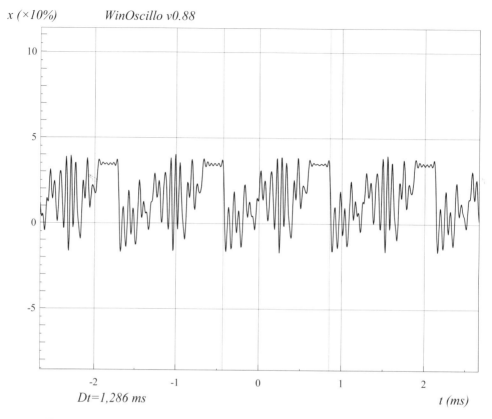

Figure 162. Bus overflow: the message "LED on" is repeated endlessly.

The solution is to use two breadboards. Each of them must carry the four decoupling capacitors on the corners as discussed in section 2.1.

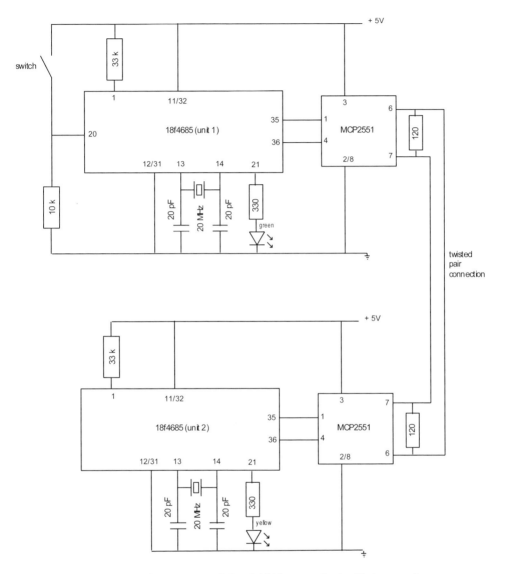

Figure 163. Schematic of the CAN bus project with two nodes.

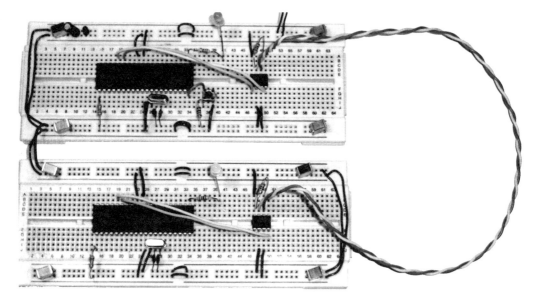

Figure 164. CAN bus project on two breadboards.

Software

This time we will not use loopback mode but normal mode. That means that cancon must be set differently.

    cancon = 0b_00000000

We need to wait for this setting to be successfully executed.

    while (canstat & 0b_1110_0000)!=(0b_0000_0000) loop end loop

The next table shows the software in both nodes next to each other. This is very interesting because you can compare the sources easily. "Upper" and "lower" refers to the picture of the breadboard.

Node 1 (upper)	Node 2 (lower)
-- JAL 2.4j include 18f4685_bert -- project canbus node 1 (top) -- select CAN config mode	-- JAL 2.4j include 18f4685_bert -- project canbus 2 (bottom) -- select CAN config mode

CANCON = 0b_1000_0000 while (CANSTAT & 0b_1110_0000)!=(0b_1000_0000) loop end loop -- set speed to 100k bit BRGCON1 = 0b_0000_0011 BRGCON2 = 0b_1011_1111 BRGCON3 = 0b_0000_0111 -- make sure TX doesn't get a tristate CIOCON = CIOCON|0b_0010_0000 -- setup is done, switch to normal mode CANCON = 0b_00000000 while (CANSTAT & 0b_1110_0000)!=(0b_0000_0000) loop end loop -- define the pins pin_b2_direction = output pin_b3_direction = input pin_d1_direction = input pin_d2_direction = output -- define variables var byte data -- transmit priority 0 TXB0CON = 0b_0000_0000 -- transmit and receive message length 8 bytes, TXB0DLC = 0b_00001000 RXB0DLC = 0b_00001000 -- recieve valid messages with standard id in buffer 0 using filter 0 RXB0CON = 0b_00100000 forever loop	CANCON = 0b_1000_0000 while (CANSTAT & 0b_1110_0000)!=(0b_1000_0000) loop end loop -- set speed to 100k bit BRGCON1 = 0b_0000_0011 BRGCON2 = 0b_1011_1111 BRGCON3 = 0b_0000_0111 -- make sure TX doesn't get a tristate CIOCON = CIOCON|0b_0010_0000 -- setup is done, switch to normal mode CANCON = 0b_00000000 while (CANSTAT & 0b_1110_0000)!=(0b_0000_0000) loop end loop -- define the pins pin_b2_direction = output pin_b3_direction = input pin_d2_direction = output -- define variables var byte data -- transmit priority 0 TXB0CON = 0b_0000_0000 -- transmit and receive message length 8 bytes, TXB0DLC = 0b_00001000 RXB0DLC = 0b_00001000 -- recieve valid messages with standard id in buffer 0 using filter 0 RXB0CON = 0b_0010_0000 forever loop

```
-- send message
if pin_d1 then

  -- set message identifier (id)
  TXB0SIDL = 0b_00100000
  TXB0SIDH = 0b_00000000

  -- use value 25
  data = 25

  -- put this into buffer 0
  TXB0D0 = data

  -- transmit, priority level 1
  TXB0CON = 0b_00001001

while(TXB0CON&0b_0000_100)==0b_00
00_100 loop end loop

  pin_d2 = high
  delay_100ms(5)
  pin_d2 = low

    -- set message identifier (id)
  TXB0SIDL = 0b_00100000
  TXB0SIDH = 0b_00000000

  -- use value 25
  data = 15

  -- put this into buffer 0
  TXB0D0 = data

  -- transmit, priority level 1
  TXB0CON = 0b_00001001

while(TXB0CON&0b_0000_100)==0b_00
00_100 loop end loop

end if

end loop
```

```
-- wait for messages in buffer 0
while
(RXB0CON&0b_1000_0000)!=0b_1000_0
000 loop end loop

-- does this message have the right id?
if
(0b_00100000==RXB0SIDL)&(0b_000000
00==RXB0SIDH) then
  -- get the data
  data = RXB0D0

  -- send to the PC
  serial_sw_write(data)

  -- light the LED if appropriate
  if data==25 then
    pin_d2 = high
  else
    pin_d2 = low
  end if

  -- clear the flag
  RXB0CON=RXB0CON&0b_0111_1111

end if

end loop
```

9.10 SPI - Master - slave

The purpose of this project is to set up a connection between two microcontrollers using the SPI protocol. This is a very interesting protocol because sending and receiving is done simultaneously.

Technical background

SPI is an abbreviation for Serial Peripheral Interface. An SPI bus has one master and at least one slave. A separate wire, the SS (slave select[40]), is used to indicate for which slave the message is intended. The disadvantage of this is that you need an additional wire for each slave. The advantage is that addressing doesn't use time on the bus. That makes the bus faster. The SS line has reversed logic. This means that the line is normally high, and is pulled low by the master if it wants to send a message to the slave. To indicate the reversed logic SS is often written as /SS.

The SPI bus itself consists of three wires, two for data and one for the clock. The clock is controlled by the master. This ensures that all IC's on the bus use the exact same frequency. One of the data lines is meant for transport of data from the master to the slave, the other for data in the opposite direction. The SDO pin - serial data out - of the master must be connected with the SDI pin - serial data in - from the slaves, and vise versa.

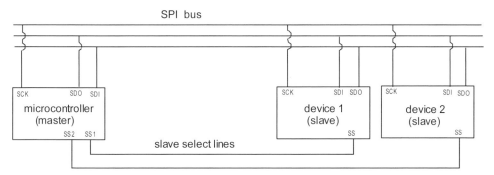

Figure 165. Structure of the SPI bus.

In the technical documentation of SPI components SPI modes are used to indicate what type of SPI bus behavior the component expects. What these modes mean is listed in the next table

---

[40] In some documentation CS, chip select, is used.

SPI mode	CKP	CKE
0,0	0	1
0,1	0	0
1,0	1	1
1,1	1	0

Table 43. SPI modes.

You see that SPI mode 0,0 does <u>not</u> mean that CKP and CKE are both zero. This is a frequent cause of errors, so beware. CKP and CKE are bits in the sspcon1 register. In the software part we will get back to this. Obviously both master and slave must support the same SPI mode, and be set accordingly, in order to be able to communicate with each other.

We will make use of the SPI unit of the microcontroller. That means we are not free to chose the pins. For the slave /SS is set (on a5), the master can chose freely. We will use pin d3 as /SS for the master.

	16F877A	18F4455	18F4685
SDO	c5	c7	c5
SDI SDA	c4	b0	c4
SCK - SCL	c3	b1	c3
/SS (slave)	a5	a5	a5

Table 44. SPI connections for three microcontrollers.

In this project we use the 16f877A microcontroller. The SPI connections are shown in the previous table, along with the connections for the other two microcontrollers that we use in this book, the 18f4455 and the 18f4685. The pins for the 18f4455 deviate from the others because it doesn't have a pin c3 - which is the V_{USB} pin - and c4 and c5 can only be input. When you define the direction of the pin please note the following:

1. The clock is an output on the master but an input on the slaves.
2. The /SS pin on the slave is a5, an analog pin that must be made digital first.

Hardware

Not visible in the next Figures are the power wires that connect all rails with each other, and the decoupling capacitors on the ends of each rail, as discussed in section 2.1

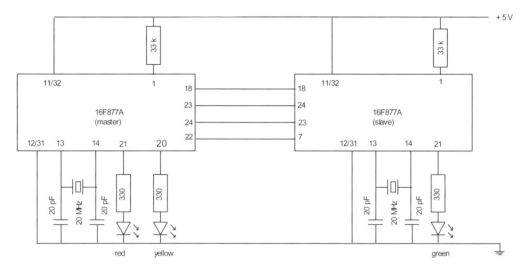

Figure 166. Schematic of the SPI mater - slave project.

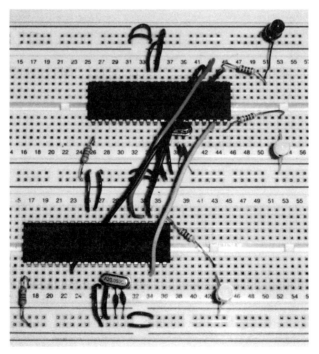

Figure 167. The master-slave project on a double breadboard.

Software

The base of the software is the setting of the SPI registers. The settings of the master must of course match the settings of the slave. In this case that is rather easy because we can make any choice we want since both master and slave are microcontrollers. Normally however you will use devices as slave that have factory configuration settings. In that case you must carefully study the datasheet of that device, and set the master accordingly. The SPI protocol is fast, but not very forgiving. A tiny error in the settings will cause communication to fail.

SSPSTAT: MSSP STATUS REGISTER (SPI MODE)

R/W-0	R/W-0	R-0	R-0	R-0	R-0	R-0	R-0
SMP	CKE[1]	D/$\overline{A}$	P	S	R/$\overline{W}$	UA	BF
bit 7							bit 0

Legend:
R = Readable bit   W = Writable bit   U = Unimplemented bit, read as '0'
-n = Value at POR   '1' = Bit is set   '0' = Bit is cleared   x = Bit is unknown

bit 7   **SMP**: Sample bit
         SPI Master mode:
         1 = Input data sampled at end of data output time
         0 = Input data sampled at middle of data output time
         SPI Slave mode:
         SMP must be cleared when SPI is used in Slave mode.

bit 6   **CKE**: SPI Clock Select bit[1]
         1 = Transmit occurs on transition from active to Idle clock state
         0 = Transmit occurs on transition from Idle to active clock state

bit 5   **D/$\overline{A}$**: Data/Address bit
         Used in I^{2}C mode only.

bit 4   **P**: Stop bit
         Used in I^{2}C mode only. This bit is cleared when the MSSP module is disabled, SSPEN is cleared.

bit 3   **S**: Start bit
         Used in I^{2}C mode only.

bit 2   **R/$\overline{W}$**: Read/Write Information bit
         Used in I^{2}C mode only.

bit 1   **UA**: Update Address bit
         Used in I^{2}C mode only.

bit 0   **BF**: Buffer Full Status bit (Receive mode only)
         1 = Receive complete, SSPBUF is full
         0 = Receive not complete, SSPBUF is empty

Note 1: Polarity of clock state is set by the CKP bit (SSPCON1<4>).

Figure 168. The sspstat register.

The registers that we need can be used for SPI as well as I²C, but have a different meaning in each case. So make sure to select the correct versions in the datasheet! In this book we have of course done that for you. It is particularly important that we make the same settings in both microcontrollers. At the end of this project both programs (for master and slave) are listed next to each other for easy comparison. We select these settings more or less arbitrarily:

1. Input data sampled at middle of data output time.
2. Transmit occurs on transition from active to Idle clock state.

So sspstat = 0b_0100_0000.

SSPCON1: MSSP CONTROL REGISTER 1 (SPI MODE)

R/W-0	R/W-0	R/W-0	R/W-0	R/W-0	R/W-0	R/W-0	R/W-0
WCOL	SSPOV[1]	SSPEN	CKP	SSPM3	SSPM2	SSPM1	SSPM0

bit 7   bit 0

Legend:
R = Readable bit   W = Writable bit   U = Unimplemented bit, read as '0'
-n = Value at POR   '1' = Bit is set   '0' = Bit is cleared   x = Bit is unknown

bit 7   **WCOL**: Write Collision Detect bit (Transmit mode only)
             1 = The SSPBUF register is written while it is still transmitting the previous word (must be cleared in software)
             0 = No collision

bit 6   **SSPOV**: Receive Overflow Indicator bit[1]
             SPI Slave mode:
             1 = A new byte is received while the SSPBUF register is still holding the previous data. In case of overflow, the data in SSPSR is lost. Overflow can only occur in Slave mode. The user must read the SSPBUF, even if only transmitting data, to avoid setting overflow (must be cleared in software).
             0 = No overflow

bit 5   **SSPEN**: Master Synchronous Serial Port Enable bit
             1 = Enables serial port and configures SCK, SDO, SDI and SS as serial port pins[2]
             0 = Disables serial port and configures these pins as I/O port pins[2]

bit 4   **CKP**: Clock Polarity Select bit
             1 = Idle state for clock is a high level
             0 = Idle state for clock is a low level

bit 3-0   **SSPM3:SSPM0**: Master Synchronous Serial Port Mode Select bits
             0101 = SPI Slave mode, clock = SCK pin. SS pin control disabled, SS can be used as I/O pin[3]
             0100 = SPI Slave mode, clock = SCK pin. SS pin control enabled[3]
             0011 = SPI Master mode, clock = TMR2 output/2[3]
             0010 = SPI Master mode, clock = Fosc/64[3]
             0001 = SPI Master mode, clock = Fosc/16[3]
             0000 = SPI Master mode, clock = Fosc/4[3]

Note 1: In Master mode, the overflow bit is not set since each new reception (and transmission) is initiated by writing to the SSPBUF register.
       2: When enabled, these pins must be properly configured as input or output.
       3: Bit combinations not specifically listed here are either reserved or implemented in I²C™ mode only.

Figure 169. The sspcon1 register.

In the sspcon1 register we need to set whether the microcontroller is a master or a slave, so here we need different settings. The first part of the register refers to the communication, so that has to be identical. We select:

1. No collision.
2. The overflow bit is not set.
3. Enable serial port and configure SCK, SDO, SDI, and SS as serial port pins.
4. Idle state for clock is high.

So sspcon1 = 0b_0011_????.

The last four bits determine what kind of unit the microcontroller is (master or slave) and the speed of the clock. When you connect SPI components to the bus you need to check what the maximum clock frequency is that they can handle. The master must never run the clock faster than that. In our case it makes no difference because both devices are identical microcontrollers. So we make a random choice, Fosc/16.

master	slave
Master mode clock = FOSC/16 sspcon1 = 0b_0011_0001	Slave mode clock = SCK pin SS pin control enabled sspcon1 = 0b_0011_0100

Since the slave cannot do anything on its own the master has full control. Since sending and receiving happens simultaneously the master has to send something, anything, in order to receive an answer. That is what makes this protocol so very interesting. Let's take a look at the PSI library.

```
function spi_transceive (byte in data) return byte is
   - send and receive bytes using the hardware SPI unit
   SSPBUF = data
   -- wait for the sending to complete
   while (!BF)  = loop end loop
   -- receive at the same time
   return SSPBUF
end function
```

The SPI hardware library consists of a single procedure. It works as follows. The data to be sent is put into the SPI buffer sspbuf. If the microcontroller is the master then sending starts automatically. For each bit that is put on the SDO (data out) line a bit is read on the SDI (data in) line, and this bit is put in the buffer. Then the next bit from the buffer is sent and received, and so on. After 8 bits the buffer is completely replaced by the incoming data. At that moment flag BF is set by the microcontroller. So the procedure waits for this flag to be set, and then copies the received data from the buffer.

The slave uses the exact same procedure. The slave can also put data in the SPI buffer sspbuf. But unlike the master it is not sent automatically. The slave will wait for the signal from the master and then use the same routine to exchange incoming and outgoing bits. A fascinating protocol.

command	description
data0 = spi_transceive(data1)	A function that sends data1 using the SPI protocol and receives data0 at the same time. Can be used in both slave and master.

This is what the two programs will do:

1. The master will flash the red LED with a frequency of 1 Hz.
2. It will transmit the status of the LED to the slave.
3. The slave will read the status and send it to the green LED.
4. At the same time the slave will send the previously received status back.
5. The master reads that, and sends it to the yellow LED.

Can you predict what the flashing will look like?

The result is that the green and red LEDs flash simultaneously, and the yellow LED exactly opposite. The yellow LED is out of synch with the two other LEDs because the slave sends the last received status back to the master. The last status is always the exact opposite of the current status. If the slave receives the message "LED on" than it just sent the previous message "LED off" back.

The /SS pin on the slave, pin a5, is an analog pin that needs to be made digital first using the adcon1 register.

ADCON1 REGISTER (ADDRESS 9Fh)

R/W-0	R/W-0	U-0	U-0	R/W-0	R/W-0	R/W-0	R/W-0
ADFM	ADCS2	—	—	PCFG3	PCFG2	PCFG1	PCFG0
bit 7							bit 0

bit 7   **ADFM**: A/D Result Format Select bit
1 = Right justified. Six (6) Most Significant bits of ADRESH are read as '0'.
0 = Left justified. Six (6) Least Significant bits of ADRESL are read as '0'.

bit 6   **ADCS2**: A/D Conversion Clock Select bit (ADCON1 bits in shaded area and in **bold**)

ADCON1 <ADCS2>	ADCON0 <ADCS1:ADCS0>	Clock Conversion
0	00	Fosc/2
0	01	Fosc/8
0	10	Fosc/32
0	11	F_{RC} (clock derived from the internal A/D RC oscillator)
1	00	Fosc/4
1	01	Fosc/16
1	10	Fosc/64
1	11	F_{RC} (clock derived from the internal A/D RC oscillator)

bit 5-4   **Unimplemented**: Read as '0'

bit 3-0   **PCFG3:PCFG0**: A/D Port Configuration Control bits

PCFG <3:0>	AN7	AN6	AN5	AN4	AN3	AN2	AN1	AN0	VREF+	VREF-	C/R
0000	A	A	A	A	A	A	A	A	VDD	VSS	8/0
0001	A	A	A	A	VREF+	A	A	A	AN3	VSS	7/1
0010	D	D	D	A	A	A	A	A	VDD	VSS	5/0
0011	D	D	D	A	VREF+	A	A	A	AN3	VSS	4/1
0100	D	D	D	D	A	D	A	A	VDD	VSS	3/0
0101	D	D	D	D	VREF+	D	A	A	AN3	VSS	2/1
011x	D	D	D	D	D	D	D	D	—	—	0/0
1000	A	A	A	A	VREF+	VREF-	A	A	AN3	AN2	6/2
1001	D	D	A	A	A	A	A	A	VDD	VSS	6/0
1010	D	D	A	A	VREF+	A	A	A	AN3	VSS	5/1
1011	D	D	A	A	VREF+	VREF-	A	A	AN3	AN2	4/2
1100	D	D	D	A	VREF+	VREF-	A	A	AN3	AN2	3/2
1101	D	D	D	D	VREF+	VREF-	A	A	AN3	AN2	2/2
1110	D	D	D	D	D	D	D	A	VDD	VSS	1/0
1111	D	D	D	D	VREF+	VREF-	D	A	AN3	AN2	1/2

A = Analog input   D = Digital I/O
C/R = # of analog input channels/# of A/D voltage references

Figure 170. The adcon1 register of the 16f877A.

Pin a5 is connected to A/D channel AN4. As you can see in the previous Figure 0100 for bit3:0 takes care of that. Since we are not using any of the analog channels in this project anyway we will not bother with the other settings in this register and simply make them zero. Which incidentally is the correct setting for bit 7 and 6 for normal analog operation as set by the _bert library.

adcon1 = 0b_0000_0100

The remainder of the programs contains no surprises. Note that the master controls the slave select line (/SS) and that this line is normally high, unless the master wants to communicate.

master	slave

`-- JAL 2.4j`
`include 16f877a_bert`
`include spi_hardware`
 	
`-- red led`
`pin_d2_direction = output`
`var bit redled is pin_d2`
`redled = low`
 	
`-- yellow led`
`pin_d1_direction = output`
`var bit yellowled is pin_d1`
`yellowled = low`
 	
`-- chip select`
`pin_d3_direction = output`
`var bit ss is pin_d3`
`ss = high`
 	
`-- other SPI pins`
`pin_c3_direction = output   -- SCK`
`pin_c4_direction = input    -- SDI`
`pin_c5_direction = output   --SDO`
 	
`-- register settings`
`sspstat = 0b_0100_0000`
`sspcon1 = 0b_0011_0001`
	`-- JAL 2.4i`
	`include 16f877a_bert`
	`include spi_hardware`
	
	`-- green led`
	`pin_d2_direction = output`
	`var bit led is pin_d2`
	`led = low`
	
	`-- chip select (input)`
	`adcon1 = 0b_0000_0100`
	`pin_a5_direction = input`
	`var bit ss is pin_a5`
	
	`-- other SPI pins`
	`pin_c3_direction = input    -- SCK (input)`
	`pin_c4_direction = input    -- SDI`
	`pin_c5_direction = output   --SDO`
	
	`-- register settings`
	`sspstat = 0b_0100_0000`
	`sspcon1 = 0b_0011_0100`

forever loop -- leds on (local and remote) redled = high ss = low yellowled = spi_transceive(1) ss = high delay_1s(1) --leds off (local and remote) redled = low ss = low yellowled = spi_transceive(0) ss = high delay_1s(1) end loop	-- variables in use var byte data forever loop -- read the data and control the green led -- and send the previous command back led = spi_transceive(data) data = led end loop

Table 45. SPI master and slave programs next to each other.

9.11 SPI - Sampling to an MMC card

The purpose of this project is to collect samples, in this case A/D measurements of a potmeter, and store them on an MMC card. We will use Michael Dworkin's routines for MMC and SPI that have been converted to JAL.[41]

Technical background

This project is meant for MMC cards and not SD cards. MMC cards have 7 contacts on the backside. The next Figure shows the connections, as seen from the side that the connections are on.

---

[41] C-routines and other information in German at http://www.cc5x.de/MMC/

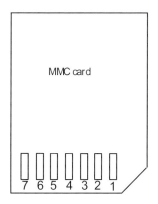

Figure 171. Connections of the MMC card.

The contacts have the following purpose:

contact	name	purpose
1	/CS	chip select
2	SDI	data in
3	GND	0 V
4	V_{DD}	+3.3 V
5	CLK	clock
6	GND	0 V
7	SDO	data out

Table 46. Connections of the MMC card.

You can purchase a special connector for the MMC card, but if you happen to have an old circuit board plug that will work just as well. The first pin of the plug cannot be used because the MMC card cannot reach it, so start soldering at the second pin. When you insert the MMC card make sure the spring contacts match up exactly with the contacts on the cards, so don't push it in all the way. If your plug has contacts on top and bottom also make sure you select the right row.

Figure 172. Make sure the spring contacts match with the card contacts.

Hardware

The MMC card needs 3.3 volts V_{DD}. That means that the voltage needs to be reduced from 5 to 3.3 volts. We will use an LM317 voltage regulator. The 1k LIN potmeter is used to adjust the voltage. When you are done with the hardware connect the entire project but <u>without the MMC card</u>. Then use a voltage meter to adjust the voltage of the regulator to 3.3 volts.

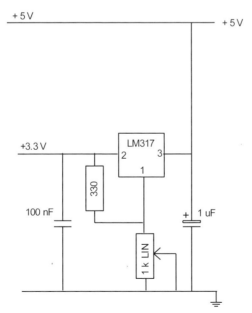

Figure 173. MMC card power supply.

The inputs of the MMC card need 3.3 volts as well, which means we cannot connect it directly to the microcontroller. Instead we will use a voltage divider on the output of the microcontroller. This divider consists or two resistors of 3k3 and 1k8 in series. The current from the microcontroller through both resistors when the pin is high is:

$$I = V/R = 5/(3k3+1k8) = 1 \text{ mA}$$

Far below the maximum value of 25 mA (and 200 mA per port). The resulting voltage between the two resistors at that current - so when the pin is high - is:

$$V = 5 * (3k3/(3k3+1k8)) = 3.24 \text{ V}$$

This is close enough. The card outputs are of course also 3.3 volts, but that is high enough to be regarded as a "1" by the microcontroller. According to the datasheet the minimum "1" value is 2 volts. That means we do not need to convert that signal.

It is by the way perfectly possible to run the entire microcontroller at 3.3 volts, but that reduces the A/D range as well. Besides the passthrough functionality of the Wisp wouldn't work anymore. For that reason we have opted for a dual voltage setup.

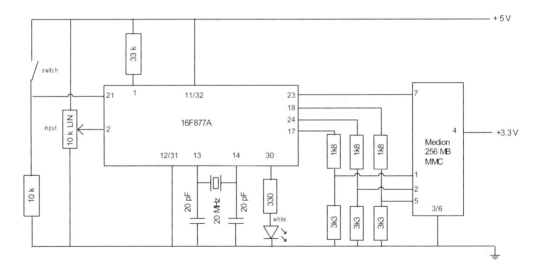

Figure 174. Schematic of the MMC card project, without 3.3 volts regulator.

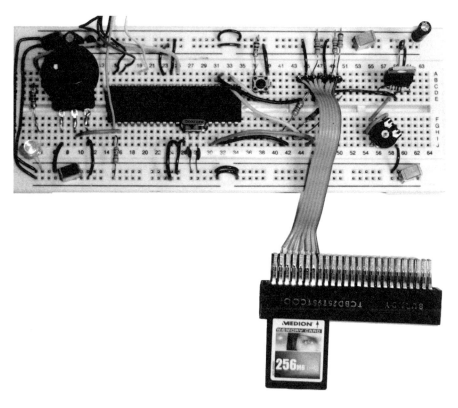

Figure 175. The project on a breadboard, including the 3.3 volts regulator.

Software

As you have probably noticed the MMC card uses SPI for communication. In this project we will use a software SPI library which allows you to use this project also on microcontrollers that do not have SPI functionality built in. The master clock frequency is set on about 620 kHz, a good value for controlling the MMC card.

command	description
data0 = spi_transceive(data1)	A function that sends data1 using the SPI protocol and receives data0 at the same time. Can be used in both slave and master.

Of course you can use the hardware SPI library if you want. Bear in mind that in that case you may need to change the connections, and that you must set the SPI registers in the master correctly.

We will use a special MMC library. This library assumes that an SPI library is loaded first, and contains the following functionality.

command	description
mmc(byte in Cmd,word in dataH,word in dataL,byte in CRC)	Send a command to the MMC card with dataH and dataL as payload, including an CRC.
mmc_init	Initialize the MMC card.
MMC_write_open(word in dataH, word in dataL)	Open the MMC card for a block write operation (in blocks of 512 bytes), with dataH and dataL as payload.
MMC_write_close	Close the block write operation.
MMC_read_open(word out dataH, word out dataL)	Open the MMC card for a block read operation (in blocks of 512 bytes), with dataH and dataL as payload.
MMC_read_close	Close the block read operation.
MMC_read_streaming_open(word in dataH,word in dataL)	Open the MMC card in streaming mode, with dataH and dataL as payload.
MMC_read_streaming_close	Close the streaming mode.
Possible error messages generated by the functions (use for example error = MMC_write_open(0,0)) 0 = everything ok 1 = block write open failed 2 = block write failed 3 = block read open failed 4 = streaming mode open failed	

Using these commands writing the program is not very complicated.

```
--JAL 2.4j
include 16F877A_bert
include spi_software
include mmc

-- pins
pin_a0_direction = input   -- var resist
pin_d2_direction = input   -- switch
var bit switch is pin_d2
pin_d7_direction = output  -- LED
var bit led is pin_d7

-- variables
var byte resist = 0

-- wait here for the button to be pressed
led = 1
delay_1s(1)
led = 0
while !switch loop end loop

-- INIT the MCC card
MMC_Init

-- WRITE data in 512 byte mode at address 0,0
MMC_write_open(0,0)
-- measure and save
led = 1
for 512 loop
   resist = adc_read_low_res(0)
   spi_transceive(resist)
   delay_100ms(1)
end loop
led = 0
-- close the write operation
MMC_write_close

-- READ data in 512 byte mode at address 0,0
MMC_read_open(0,0)
for 512 loop
   -- read and send to the PC
   serial_sw_write(spi_transceive(0xFF))
```

```
        end loop
        MMC_read_close

        -- Done flash the LED
        forever loop
           led = on
           delay_100ms(1)
           led = off
           delay_100ms(1)
        end loop
```

Instructions

1. Start the program, and start MICterm with the Wisp628/648 selected. Connect to the microcontroller at 1200 baud, and select as data "RAW".
2. Wait for the LED to flash shortly, and then press the button.
3. The program will now take 512 samples and store them on the MMC card. How long that takes depends on the delay between two samples. If the delay is 100 mS it will take 51.2 seconds, but if the delay is 1 second it will take 8.5 minutes. In the latter case it will take 7.9 years for the entire 256 MB card to fill up. The LED is on while sampling is in progress. You need to wait for the LED to go off before you can stop the program - assuming that you do not want to corrupt the data.
4. When the samples are collected and stored on the card they are read back and send to the PC - MICterm in this case - for viewing and analyses. In a real application you will sample and store "in the field" and read and analyze the data when you get back to your office.
5. The LED will flash to indicate that the last data has been sent to the PC and the program is now terminated.

If you want to modify this program it is important to know that MMC cards store data in blocks of 512 bytes. If you have less data to store you still need to write 512 bytes, so fill the block with dummy data, such as zeros. It would probably be a very good idea to make a button that will stop sampling, and adds dummy data to the MMC card to finish the last block.

9.12 I^{2}C - Real Time Clock (RTC)

I^{2}C is an acronym for Inter IC Communication, and is often also abbreviated to I2C. It is a serial protocol developed by Philips in the eighties for the communication between IC's inside a single machine.

The I²C bus uses two wires, one for data and one for the clock. The clock is controlled by the master and can have a maximum speed of 100 kbit per second. Remarkable about the structure of the bus is that it requires a pull up resistor for both wires.

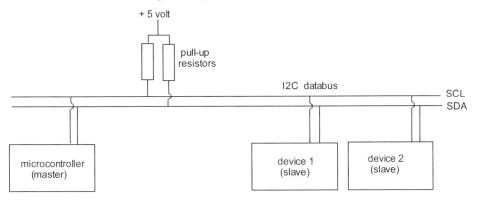

Figure 176. Structure of the I²C bus.

Normally speaking an I²C bus has as single master and one or more slaves. The master addresses the slaves using an address and takes care of the clock. The slaves can influence the clock because they can pull the clock line down preventing the master from issuing another clock pulse. This will happen for example when the master asks a question to a slave and the slave needs some time to collect the answer and prepare it for transmission. In this project the master is a microcontroller and the slave an RTC (real time clock) chip, a special clock IC.

Technical background

We will use a 16f877A microcontroller with a DS1307 real time clock manufactured by Dallas Semiconductor. The DS1307 is an I²C chip. The address is hidden in the datasheet but turns out to be 0b_1101_000r/w.

	BIT7							BIT0	
00H	CH	\multicolumn{4}{l}{10 SECONDS}	\multicolumn{3}{l}{SECONDS}	00–59
	()	\multicolumn{4}{l}{10 MINUTES}	\multicolumn{3}{l}{MINUTES}	00–59
	()	12 / 24	10 HR / A/P	10 HR		\multicolumn{3}{l}{HOURS}	01–12 / 00–23
	()	()	()	()	()	\multicolumn{3}{l}{DAY}	1–7
	()	()	\multicolumn{3}{l}{10 DATE}	\multicolumn{3}{l}{DATE}	01–28/29 / 01–30 / 01–31
	()	()	()	\multicolumn{2}{l}{10 MONTH}	\multicolumn{3}{l}{MONTH}	01–12
	\multicolumn{5}{l}{10 YEAR}	\multicolumn{3}{l}{YEAR}	00–99
07H	OUT	()	()	SQWE	()	()	RS1	RS0	

Figure 177. DS1307 time registers.

First the oscillator (CH bit 0) must be switched on because the default position is off. That is a write command so we follow the instructions in the datasheet for "writing". The commands that we discuss here are in a special library for the DS1307. You can include and use that library without understanding the commands, so if you want you can skip this part of the background. Each slave however has its own rules. So it is very convenient to understand how communication is set up based on a datasheet in case you need to do this for a component that has no JAL library. In that case you can simply make one yourself.

We will use the basic I²C library. This is a software library that can also be used in microcontrollers that do not have I²C capabilities. The pins that this library uses are in a separate file called i2cp.jal in the library directory.

command	description
i2c_put_write_address(x)	The master announces that it wants to write to the slave with address x.
i2c_put_data(x)	Send data x.
i2c_put_stop	Send a stop signal.
i2c_put_read_address(x)	The master announces that it wants to receive data from the slave with address x.
i2c_get_data(x)	Receive data.
i2c_put_ack	Send an ack (acknowledge).
i2c_put_nack	Send a nack (not acknowledge).
i2c_wait_ack	Wait for an ack (acknowledge).

According to the datasheet the write routine of the DS1307 should be as follows:

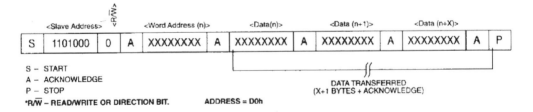

Figure 178. Writing data to a DS1307 slave.

In the time registers Figure (a few pages back) you can see that the oscillator (CH) is bit 7 of address 0. To enable the oscillator this bit must be cleared (made zero). Please note: every time you do this the clock is reset so if you want the clock to keep running correctly when you switch the microcontroller off and then back on you should not issue this command as a default.

We can send a maximum of three data bytes. Less is also possible by sending a "stop" command after receiving the acknowledge. Note that the address has only 7 bits. This is because the read/write bit needs to be added to it by the read or write routine. Once that has been added we have 8 bits. Let's follow the steps in the previous Figure.

command	description
i2c_put_write_address(0b_1101_000)	Tell the slave with address 0b_1101_000 (the 8th bit will be added automatically) that we wish to write...
i2c_put_data(0)	... to register zero.
i2c_wait_ack	Wait for the acknowledgement from the slave that the message has been received.
i2c_put_data(0)	Send the data that we wish to put in register zero (a zero).
i2c_wait_ack	Wait for the acknowledgement of the slave that the message has been received.
i2c_put_stop	Tell the slave that we will not be sending more data.

Using this same technique the time can be entered into the RTC. We will send a message to the RTC with the address and the "write" bit, followed by the address of the first register we want to write to, and then three data bytes. With each new data byte the register will automatically be incremented. In this program we will not set the seconds but enter a zero in that register instead. Technically we could have started one address higher but somehow this feels more appropriate, and we are now certain that any content is cleared. We are sending data so we must wait for an acknowledgement before we can send some more. And we terminate again with "stop" command.

command	description
i2c_put_write_address(0b_1101_000)	Tell to the slave with address 0b_1101_000 that we wish to write...
i2c_put_data(0)	... to register zero.
i2c_wait_ack	Wait for the acknowledgement from the slave that the message has been received.
i2c_put_data(0)	Send a zero for the seconds (note that this also resets the clock for CH is in this register too)...
i2c_wait_ack	...and wait for the acknowledgement.
i2c_put_data(minutes)	Send the minutes...
i2c_wait_ack	...and wait for the acknowledgement.
i2c_put_data(hours)	Send the hours...
i2c_wait_ack	...and wait for the acknowledgement.
i2c_put_stop	Tell the slave that we will not be sending more data.

The data are distributed over the registers in a rather inconvenient way. We need to shift the decimals left in order to fit the units next to them. The variables sm, tm, sh, th will be entered using a serial RS232 connection (you will be able to set the time using your PC) and are thus in ASCII. In order to use them we need to deduct 48 to get from ASCII to normal numbers:

    seconds = 0
    minutes = 0
    minutes = ((tm-48)<<4)+(sm-48)
    hours = 0
    hours = ((th-48)<<4) + (sh-48)

Reading will take place from the last used register. So first we write to the clock to indicate that we want to start with register zero.

    i2c_put_write_address(0b_01101_000)
    i2c_put_data(0b_0000_000)
    i2c_wait_ack
    i2c_put_stop

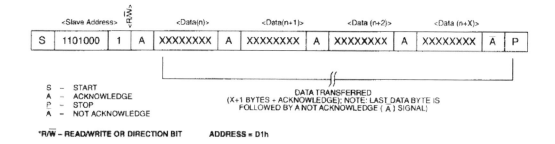

Figure 179. Reading data from the DS1307 slave.

The reading routine must be executed according to the previous Figure from the RTC datasheet. In the reading routine "we" are receiving data, so "we" must confirm the acknowledgement to the slave. That means we need to use a "put_ack" command instead of a "wait_ack" command. After receipt of the time we terminate communications using a "nack" and a "stop" command, exactly as prescribed.

    i2c_put_read_address(0b_01101_000)
    i2c_get_data(seconds)
    i2c_put_ack
    i2c_get_data(minutes)
    i2c_put_ack
    i2c_get_data(hours)
    i2c_put_nack
    i2c_put_stop

These are all the basic routines. The only thing left to do is put these in a separate file as functions, and give the file a suitable name. This has already been done for you. You will find the procedures in the DS1307 library.

command	description
ds1307_entertime(sec,min,hr)	Request the time using a PC screen and prepare the answers for the next command.
ds1307_sendtime(sec,min,hr)	Send the time to the ds1307.
ds1307_readtime	Request the time from the DS1307 and prepare the answers for the next command.
ds1307_showtime	Display the time on the PC screen using the format hh:mm:ss

Hardware

As microcontroller we have opted for the 16f877A, powered by the Wisp648 (or another 5 volts power supply). The DS1307 is powered by the Wisp too, but also has a backup battery. When the microcontroller is powered down the clock will keep running on its battery. Hence the name real time clock. As battery a 3 volts lithium button cell is used with two wires soldered to it. You have to solder the battery quickly to keep it as cool as possible. Of course using a proper battery holder is probably a better idea.

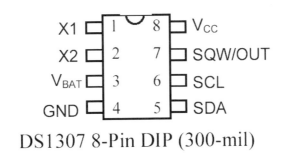

Figure 180. Pin layout of the DS1307 real time clock.

For the I²C wires 2k2 (red-red-red) pull up resistors are used. The DS1307 uses its own clock crystal with a special frequency of 32.768 kHz. This frequency is often used in clocks and watches. The microcontroller could generate this frequency for the RTC, but then the clock would stop if the microcontroller stops, which is quite the opposite of what we want to achieve. The clock crystal must be connected without capacitors.

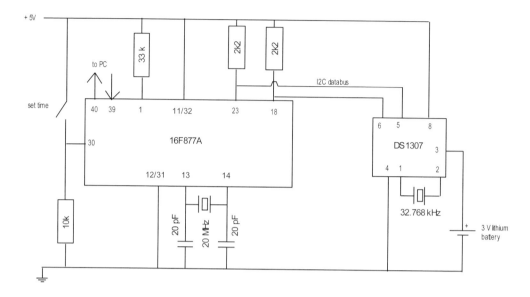

Figure 181. Real time clock schematic.

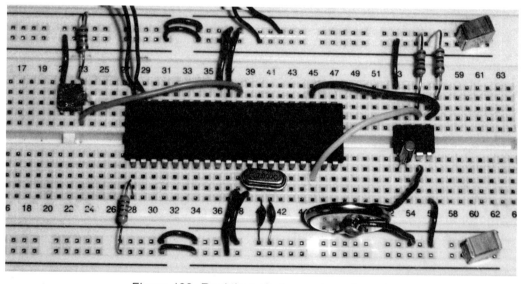

Figure 182. Real time clock on a breadboard.

Software

The content of the software has already been discussed in the technical background section. When you connect the clock to HyperTerminal (at 1200 baud) you can immediately see the running time. Please remember that HyperTerm is not capable of switching the Wisp into passthrough mode, so you need the program WispPassThrough for that. When this program is started you will get a small window that always stays on top.

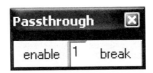

Figure 183. Passthrough window.

If you click on "enable" the Wisp is put in the passthrough mode, and the signal will go from red to green. With "break" you can take the Wisp out of passthrough mode. Normally speaking this is never necessary because when programming starts, or the power is disconnected, the Wisp will disable passthrough mode automatically. Make sure that the COM port is free before enabling passthrough mode. So first enable passthrough and only then connect using the communication package. If the port is not free the signal will go from red to yellow but not to green.

To set the clock press the button on the breadboard and follow the instructions on the screen. You cannot correct mistakes using the backspace key. If you make a mistake continue with dummy numbers, and then press the button again to re-try.

```
-- JAL 2.4i
include 16F877A_bert
include i2c
include ds1307

-- variables
var byte sec,min, hrs

-- button
pin_d7_direction = input

-- cursor home
serial_sw_write(27)
serial_sw_write("H")

-- clear the screen
serial_sw_write(27)
```

```
serial_sw_write("J")

forever loop

   -- if the button is engaged the time must be set
   if pin_d7 then

      -- cursor home
      serial_sw_write(27)
      serial_sw_write("H")

      -- clear the screen
      serial_sw_write(27)
      serial_sw_write("J")

      -- enter the starting time on a VT52 terminal
      ds1307_entertime(sec, min, hrs)

      -- and send it to the clock
      ds1307_sendtime(sec, min, hrs)

      -- cursor home
      serial_sw_write(27)
      serial_sw_write("H")

      -- clear the screen
      serial_sw_write(27)
      serial_sw_write("J")
   end if

   -- read the time
   ds1307_readtime(sec, min, hrs)
   -- and display it on a PC
   ds1307_showtime(sec,min,hrs)

end loop
```

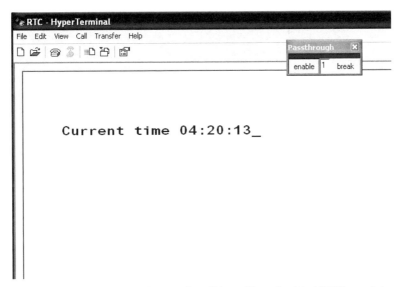

Figure 184. The clock is running (HyperTerminal in VT52 mode).

9.13 I²C - Egg timer

In this project we will use the real time clock to make an egg timer.

Technical background

The concept of an egg timer is that it needs to sound an alarm after a certain time, and that you need to be able to reset it even if the time has not expired yet. That means that the program must continuously check a button to see if the user pressed it. This time however we will use a totally different approach. We normally connect pin 1 with a 33k pull-up resistor to the +5 volts. This pin can also be used as a reset pin, by making it low. The pull-up resistor is there to prevent this from happening accidentally. Now the program doesn't have to keep an eye on the button. If it is pressed the microcontroller will reset thus causing the program to restart, presumably to the section where you can enter the time.

Hardware

In this project we will use a 16f877A microcontroller and a DS1307 real time clock IC. The DS1307 uses its own clock crystal with a special frequency of 32.768 kHz.

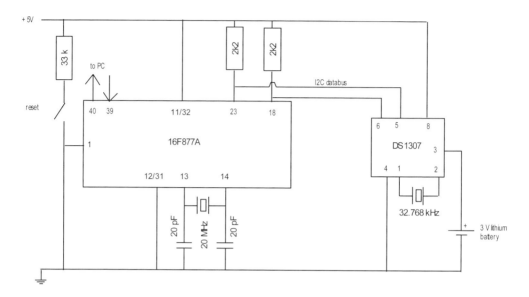

Figure 185. Schematic of the egg timer.

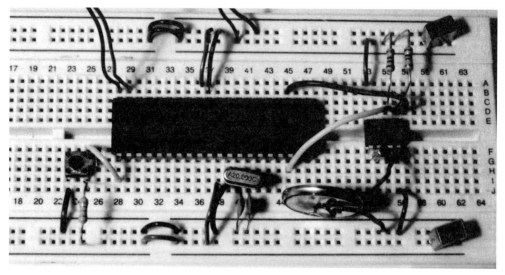

Figure 186. Egg timer on a breadboard.

Software

It would be very convenient if the RTC could run backwards, but unfortunately that is not the case. So these are the steps we will take in the program:

1. Ask for the time (ignore the hours, if any).
2. Convert this to a number of seconds.
3. Start the clock at 00:00:00.
4. Use a loop to request the time from the RTC.
5. Calculate the number of seconds that have passed (since 00:00:00).
6. Compare to the set time, and show the difference.
7. If the difference is zero the egg is ready.

In principle this is very simple, but the shifting, conversion from ASCII to normal and from time to seconds et cetera make it a bit more complicated than it sounds. Let's look at the steps again, but this time with JAL code.

1. Ask for the time (ignore the hours, if any).
2. Convert this to a number of seconds. Because we make use of the standard procedure ds1307_entertime(sec, min, hrs) from the DS1307 library we need to shift the answers around a bit in order to calculate the target number of seconds it is a bit difficult to follow. If you do this calculation on paper, using binary numbers, it is much easier to follow. This is what I do when things get complicated. Perhaps it is convenient to use the AND (&) truth table

number 1	number 2	number 1 & number 2
1	1	1 & 1 = 1
1	0	1 & 0 = 0
0	1	0 & 1 = 0
0	0	0 & 0 = 0

Table 47. AND truth table.

```
-- calculate target seconds
sectarget = (min & 0b_0111_0000) >> 4
sectarget = sectarget * 10
sectarget = sectarget + (min & 0b_0000_1111)
sectarget = sectarget * 60
sectarget = sectarget + ((sec & 0b_0111_0000) >> 4 ) * 10
sectarget = sectarget + (sec & 0b_0000_1111)
```

8. Start the clock at 00:00:00.
9. Use a loop to request the time from the RTC.
10. Calculate the number of seconds that have passed (since 00:00:00).

    ```
    -- calculate actual secondes
    secactual = (min & 0b_0111_0000) >> 4
    secactual = secactual * 10
    secactual = secactual + (min & 0b_0000_1111)
    secactual = secactual * 60
    secactual = secactual + ((sec & 0b_0111_0000) >> 4) * 10
    secactual = secactual + (sec & 0b_0000_1111)
    ```

11. Compare to the set time, and show the difference.

    ```
    -- calculate the remaing time
    if sectarget > secactual then
        secremain = sectarget-secactual
    else
        secremain = 0
    end if
    ```

12. If the difference is zero the egg is ready.

The rest of the program doesn't contain any surprises and is therefor not printed in the book. The full source code is part of the free download package.

9.14 I²C - Memory with a back-up battery

The DS1307 RTC has, besides the clock, also 56 bytes of RAM memory, which is protected by the backup battery. That means you can use this to store data in a safe place in case the power goes down. Of course you can do the same thing with EEPROM memory, but this adds another 56 bytes. And it will be safe even if you reprogram the microcontroller, or exchange it for another one.

Technical background

On top of the DS1307 memory you will find a small section of RAM memory of 56 x 8 bits, so in fact simply 56 bytes. Writing to memory, or reading from it, is handled the same way as writing or reading for example seconds.

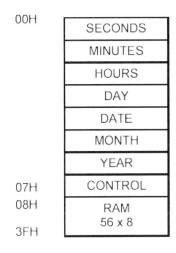

Figure 187. Memory of the DS1307.

Hardware

We will use the 16f877A microcontroller and a DS1307 real time clock IC. The DS1307 uses its own clock crystal with a special frequency of 32.768 kHz.

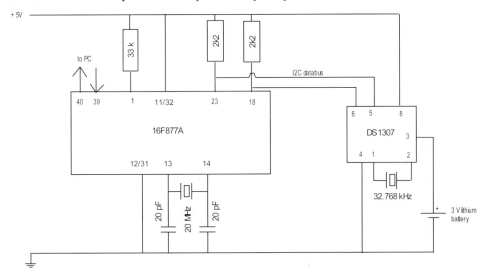

Figure 188. Schematic of the battery protected memory.

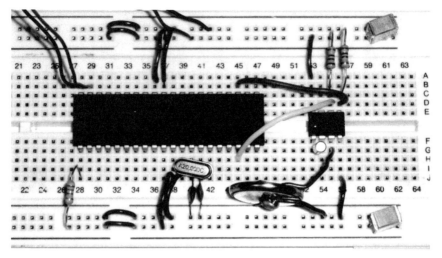

Figure 189. The battery protected memory on a breadboard.

Software

The available addresses are 0x08 up to and including 0x3FH. In this program we will use just address 0x08. The DS1307 library doesn't support this functionality, but you can easily add the procedures yourself.

```
-- JAL 2.4i
include 16F877A_bert
include i2c

-- variables
var byte data

-- point to the register we want to start reading
i2c_put_write_address( 0b_01101_000 )
i2c_put_data(0x08)
i2c_wait_ack
i2c_put_stop

-- get the data from that register
i2c_put_read_address( 0b_01101_000 )
i2c_get_data(data)
-- say we want no more bytes
i2c_put_nack
i2c_put_stop
```

forever loop

   -- increment data
   data = data + 1

   -- send the data to the clock RAM memory
   -- address
   i2c_put_write_address(0b_1101_000)
   -- register
   i2c_put_data(0x08)
   i2c_wait_ack
   -- content
   i2c_put_data(data)
   i2c_wait_ack
   -- say no more data to send
   i2c_put_stop

   -- send to the PC
   serial_sw_write(data)

   -- wait a bit
   delay_100ms(1)

end loop

Instructions

Start the program and MICterm (check Wisp628/648, 1200 baud, and select "RAW"). The numbers on the PC will count upwards. Disconnect the power by pulling the red Wisp wire from the breadboard and everything stops. Put the wire back in and the counting continues where it left off. You can reprogram the microcontroller, or swap it for another one - with the same program - and each time counting will continue where it left of.

9.15 I²C - Eight pin I/O expander

In this project we use the MCP23008 I/O expander manufactured my Microchip. This IC contains a complete port (8 pins) with complete control including interrupt on change with I²C communication to the microcontroller. If you need more inputs or outputs this is an ideal way of adding them.

Technical background

TABLE 1-2: REGISTER ADDRESSES

Address	Access to:
00h	IODIR
01h	IPOL
02h	GPINTEN
03h	DEFVAL
04h	INTCON
05h	IOCON
06h	GPPU
07h	INTF
08h	INTCAP (Read-only)
09h	GPIO
0Ah	OLAT

Figure 190. Registers of the I/O expander.

The MCP23008 is controlled using a series of registers. These registers have the following purpose:

register	description
iodir	Set a pin to input (1) or output (0). Bit 0 is for pin GP0 etc.
ipol	Set the polarity to normal (0) or reversed (1). By selecting reversed polarity a high signal on a pin will be read as low.
gpinten	Enable interrupt on change by pin (1 is on and 0 is off)
defval	Settings for use in the next register in this table.
intcon	Indicate whether the change in "interrupt on change" is compared against the previous status of the pin, or against the value in the defval register.
iocon	Configuration register, see the next Figure.
gppu	Weak pull-up per pin, makes a not connected pin high when read.
inft	Interrupt flag register to see which interrupt has occurred (read only).
intcap	Stores the status of all pins when an interrupt on change occurs (read only).
gpio	Register to make pins high or low.
olat	Shadow register of the gpio register.

IOCON – I/O EXPANDER CONFIGURATION REGISTER (ADDR 0x05)

U-0	U-0	R/W-0	R/W-0	R/W-0	R/W-0	R/W-0	U-0
—	—	SEQOP	DISSLW	HAEN	ODR	INTPOL	—
bit 7							bit 0

bit 7-6   **Unimplemented:** Read as '0'.

bit 5   **SEQOP:** Sequential Operation mode bit.
1 = Sequential operation disabled, address pointer does not increment.
0 = Sequential operation enabled, address pointer increments.

bit 4   **DISSLW:** Slew Rate control bit for SDA output.
1 = Slew rate disabled.
0 = Slew rate enabled.

bit 3   **HAEN:** Hardware Address Enable bit (MCP23S08 only).
Address pins are always enabled on MCP23008.
1 = Enables the MCP23S08 address pins.
0 = Disables the MCP23S08 address pins.

bit 2   **ODR:** This bit configures the INT pin as an open-drain output.
1 = Open-drain output (overrides the INTPOL bit).
0 = Active driver output (INTPOL bit sets the polarity).

bit 1   **INTPOL:** This bit sets the polarity of the INT output pin.
1 = Active-high.
0 = Active-low.

bit 0   **Unimplemented:** Read as '0'.

Figure 191. The iocon register of the MCP23008 I/O expander.

In this project we will connect a LED and a button to the expander. The LED should be on as long as the button is pressed. That means we need to take the following steps:

1. Set pin 0 to input and pin 2 (and why not the rest as well) to output.

   iodir = 0b_0000_0001

2. Configure the I/O expander as follows:

bit	value	description
bit 7,6	0	No function.
bit 5	0	Sequential mode.
bit 4	1	Slew rate off
bit 3	0	No function.
bit 2	0	No open drain but normal.
bit 1	1	Interrupt pin active high.
bit 0	0	No function.

So

iocon = 0b_0001_0010

3. Check the status of the button.

switch = gpio & 0b_0000_0001

4. Switch the LED on if necessary.

gpio = 0b_0000_0100

And back off.

gpio = 0b_0000_0000

A complicating factor is that these commands must be given using the I^{2}C protocol. In the hardware section you can see that we have selected address 001 for the expander by pulling pin A0 up to +5 volts and A1 and A2 down to the ground. The general address of the I/O expander must be added to that. The next Figure from the datasheet shows that the device number is 0100 so the total address becomes:

0b_0100_001x

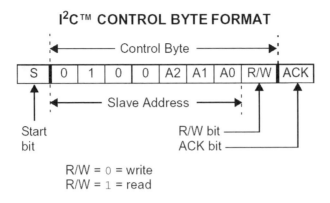

Figure 192. Address setup using pins A2:A0.

Just as in the other I²C projects we can follow the steps in the datasheet and replace them one by one with I²C commands from the I²C library. This way we can make our own MCP23008 library. The datasheet contains a wealth of information on the protocol. The part relevant for our project is shown in the next Figure.

MCP23008 I²C™ DEVICE PROTOCOL

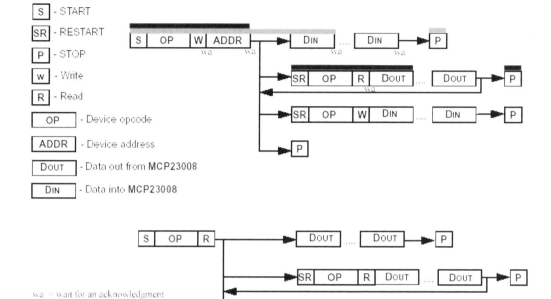

Figure 193. I²C device protocol of the MCP23008.

Using the previous Figure we can make a procedure for writing to the I/O expander - in the Figure indicated with a light gray line. Unfortunately it doesn't say when the MCP23008 will send an acknowledgement (ack) that we need to wait for. Luckily it is a bit obvious, but nevertheless it should have been included.

We will call the procedure "mcp23008_write". Since we want to put the procedure in a library we will make it a general procedure, and not hard-code the address of our I/O expander. Other users of this library may want to use a different address, or they may have multiple I/O expanders on the same bus. So this procedure needs an address where the message needs to be sent to, a register for which the data are meant, and the actual data. These are all bytes, and they need to go into the procedure.

```
procedure mcp23008_write(byte in address, byte in register, byte in data) is

    -- write data to a register in the mcp23008 i/o expander
    i2c_put_write_address( address )
    -- register
    i2c_put_data(register)
    i2c_wait_ack
    -- data for that register
    i2c_put_data(data)
    i2c_wait_ack
    -- say no more data to send
    i2c_put_stop

end procedure
```

It would be nice if we could do the opposite as well: request the content of a register on the MCP23008. Again we have two things to give to the procedure: the address of the I/O expander and the register that we are interested in. But this time we expect data back, namely the reply from the I/O expander. So that would be a byte out of the procedure, that we will call "data". In the previous Figure this procedure is indicated with a dark gray line. Note that you will get the status of all pins, also the outputs.

```
procedure mcp23008_read(byte in address, byte in register, byte out data) is

    -- read data from a register in the mcp23008 i/o expander
    i2c_put_write_address( address )
    -- register
    i2c_put_data(register)
    i2c_wait_ack
    -- switch to reading
    i2c_put_read_address( address )
    -- data from that register
    i2c_get_data(data)
    -- say no more data to get
    i2c_put_stop

end procedure
```

We will now make a new file using for example JALedit or Windows Notepad. It is better not to use a word processor for this because it may not save the file in straight ASCII but include fonts and other word processing commands. We copy both procedures into that file and save it as "mcp23008.jal" in the library directory[42]. In principle your library is now finished, but nobody knows what it is for and who wrote it. It is common practice to add some text at the top of the file that contains that information. It is also wise to state how the library may be used. Usually you can do anything you want but sometimes libraries may have restrictions. Since this library requires the use of another library we state that too. For our library we will use this preamble:

```
-- -----------------------------------------------------------------------------
-- MCP23008
--
-- Copyright: this library and the ideas, software, models, pictures contained
-- herein are Copyright (c) 1995-2010 Bert van Dam, and are distributed under
-- the Free Software Foundation General Public License version 2.0. See the
-- FSF GPL page for more information.
--
-- -----------------------------------------------------------------------------
-- Library for the mcp23008 i/o expander
-- Requirement: the i2c library
--
-- -----------------------------------------------------------------------------
-- Version: 1.0  January 19, 2010,  Bert van Dam
--  - original release
-- -----------------------------------------------------------------------------
```

Do remember to start each line with the comment indicator (--). This new library can now be used by including it in the main program. This library requires the i2c library because the procedures from that library are used in this one, so that library needs to be called first. So your program should start with:

```
include 16f877a_bert
include i2c
include mcp23008
```

Hardware

In this project we use the MCP23008 I/O expander manufactured by Microchip and the 16f877A microcontroller coincidentally also manufactured by Microchip.

---

[42] Note that I have already done this for you.

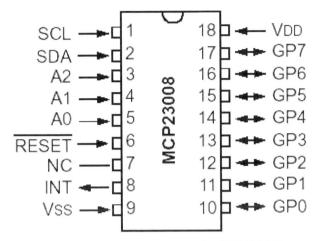

Figure 194. Pin layout of the MCP23008.

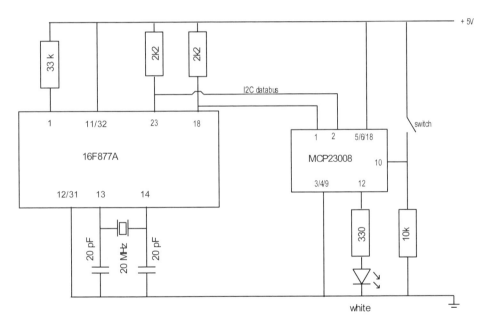

Figure 195. Schematic of the I/O expander.

The A2:A0 pins are used to set a part of the I²C address. Due to this feature you can use multiple - a maximum of eight - I/O expanders on the same bus. The selected address for this project is 001 -A2 and A1 connected to the ground, and A0 to the plus. The new port that this chip adds consists of GP0 up to and including GP7. A button is connected to GP0 and a white LED to GP2. Please note that the /reset pin is connected directly to +5 volts, without the 33 k resistor that you would normally use in the case of a microcontroller. Of course you can use any color LED you want.

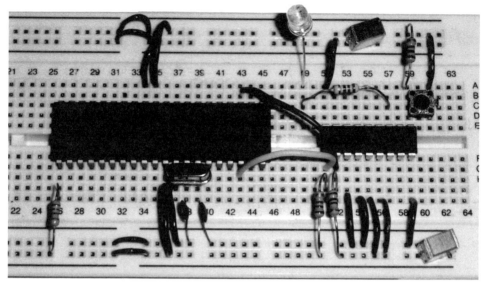

Figure 196. The project on a breadboard.

Software

In the software we will use the new commands that we have just written and stored in the MCP23008 library. So step one:

    iocon = 0b_0001_0010

Looks as follows over I^2C:

    mcp23008_write(0b_0100_001,5, 0b_0001_0010)

You can read this as: on the I^2C chip with address 0b_0100_001 the number 5 register (iocon) must contain the number 0b_0001_0010. In this way all steps from the technical background can be converted to I^2C commands, with this result:

```
-- JAL 2.4j
include 16f877a_bert
include i2c
include mcp23008
```

```
-- variables
var byte switch, tmp

-- iocon
mcp23008_write(0b_0100_001,5, 0b_0001_0010)

-- iodir
mcp23008_write(0b_0100_001,0, 0b_0000_0001)

-- read the switch and send the result to the LED
forever loop

   -- GPIO (read the switch)
   mcp23008_read(0b_0100_001,9, tmp)
   switch = tmp & 0b_0000_0001

   -- light the LED if the switch is engaged
   if switch==1 then
     -- LED on
     mcp23008_write(0b_0100_001,9, 0b_0000_0100)
   else
     -- GPIO (LED off)
     mcp23008_write(0b_0100_001,9, 0b_0000_0000)
   end if

end loop
```

The operation is simple: wait a moment for the settings to be processed and then press the button to light the LED.

9.16 I²C - D/A conversion

In this project we will use a MAX 517 D/A chip to convert a digital signal to an analog one. The analog signal can be set using two buttons, and is visualized using a LED. The other D/A project in this book - project 6.2 - uses PWM and a low pass filter instead of a special chip.

Technical background

The next Figure from the datasheet of the MAX517 shows the I²C communication protocol. There is just one type of message that can be sent to the D/A converter. This message consists of three bytes: the address of the chip, the command and a data byte.

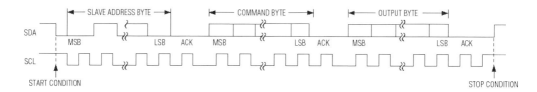

Figure 197. I²C communication protocol.

The address of the MAX517 consists of a fixed part - 010 - to indicate this type of DAC, and a variable part AD3:AD0. In the MAX517 AD3 and AD2 do not exist and read one. The two pins that this unit does have, AD1 and AD0 are pulled to the ground and thus zero, So the complete address is 0b_0101_100x

The MAX517 can be controlled with just two commands:

command	description
0	Write a value.
8	Switch the IC off.

The data byte is the output value. The full range for the data byte is 0 to 255. The maximum voltage the D/A can supply is slightly lower than the voltage of the power supply. I have measured 4.6 volts. That means each step is about 0.018 volts.

As in the previous projects the communication protocol has been converted to a procedure using the commands from the I²C library. Since there is only one command we will not turn it into a library, although it is of course possible should you want to have such a library anyway.

```
procedure max517_write(byte in address, byte in command, byte in data) is

  -- send the address (intent to write)
  i2c_put_write_address( address )
  -- register
  i2c_put_data(command)
  i2c_wait_ack
```

```
   -- data for that register
   i2c_put_data(data)
   i2c_wait_ack
   -- say no more data to send
   i2c_put_stop

end procedure
```

You are not supposed to connect components that use large currents to a D/A converter. The maximum voltage that the MAX can deliver is about 4.6 volts, and since the LED has a forward voltage of 1.3 volts the current consumption is:

$$I = V / R = (4.6-1.3) / 330 = 10 \text{ mA}.$$

This seems a bit much for a D/A converter but the chip stays cool. Optionally you can use a white LED with a 2k2 current limiting resistor. White LEDs generally need a lower current to shine brightly.

Hardware

In this project we will use the 16f877A microcontroller, and the MAX517 D/A converter, manufactured my Maxim.

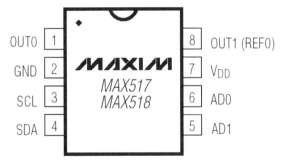

Figure 198. Pin layout of the MAX517.

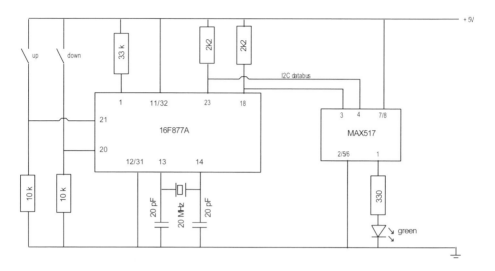

Figure 199. Schematic of the D/A converter.

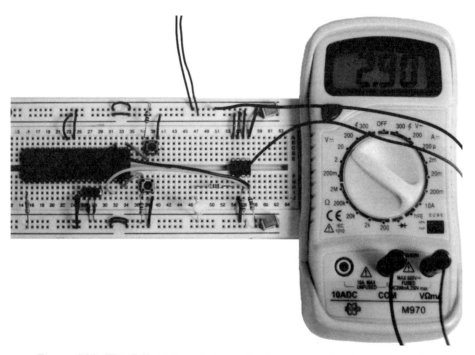

Figure 200. The D/A converter in on the breadboard with a multimeter.

In the previous Figure a multimeter is connected to pin 1, the same pin that the LED is also connected to.

Software

In the program two buttons are defined, one for up and one for down. Each time you press the button the voltage is increased or decreased in steps of 10, so with approximately 0.18 volts at a time. The program will wait for the user to let go of the button.

```
-- JAL 2.4j
include 16f877a_bert
include i2c

-- define pins and switches
pin_d1_direction = input
var bit down is pin_d1
pin_d2_direction = input
var bit up is pin_d2

-- define variables
var byte voltage = 0

procedure max517_write(byte in address, byte in command, byte in data) is

-- send the address (intent to write)
i2c_put_write_address( address )
-- register
i2c_put_data(command)
i2c_wait_ack
-- data for that register
i2c_put_data(data)
i2c_wait_ack
-- say no more data to send
i2c_put_stop

end procedure

forever loop

    if up then
```

```
   -- increment voltage
   if voltage < 250 then voltage = voltage + 10 end if

   -- wait for the button to be released
   while up loop end loop

end if


if down then
   -- decrement voltage
   if voltage > 0 then voltage = voltage - 10 end if

   -- wait for the button to be released
   while down loop end loop

end if

-- send to DAC
max517_write(0b_0101_100,0, voltage)

-- wait a bit
delay_100ms(1)


end loop
```

10 Camera Vision

In this chapter we will do a series of projects using the CMUcam2, a camera specifically designed for use in combination with a microcontroller.[43] This camera has a built in microcontroller that takes care of all the calculations. As a user you can communicate with the camera using a serial connection.

The camera needs a power supply that can deliver a minimum of 200 mA at 6 to 15 volts. That means it cannot be connected to the Wisp648 because it delivers exactly 5 volts. If you do not have another power supply you can build one yourself. The next Figure shows a variable power supply that delivers 1.2 to 13 volts. It will work perfectly with this camera, and may come in handy for other projects as well when you need a different voltage. If you use a heat sink of about 10 cm^2 this power supply can deliver a maximum of 1.5 A, more than enough.

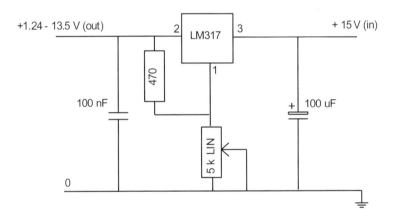

Figure 201. Schematic of the variable power supply.

Before you connect the power supply to the camera you should set it to the correct voltage using a multimeter. In all projects in this chapter I have used 8 volts for the camera. Please note: the microcontroller is connected to the Wisp648, at 5 volts.

---

[43] You can also use the CMUcam3 in compatibility mode. Please note that with the CMUcam3 camera you need to change HEX files in order to get a different baud rate. Make sure to follow the instructions that came with your camera.

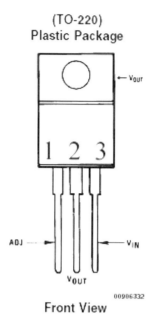

Figure 202. Pin layout of the LM317.

If you haven't used your camera yet it is highly recommended to do the tests in this preamble so you know for sure that your camera is working properly, and you get a bit of a feel for it. It is however not necessary, if you want you can skip this part and go straight to the first project.

If your PC doesn't have a serial port you need to purchase an USB-RS232 cable and use this to connect the camera to a USB port. Follow the instructions that came with the cable to find out which virtual serial port is created by the cable. If you do have a serial port than connect the camera to it, and make a note of the number. If you don't know the number assume it is port one for the time being.

You can hold the camera in a certain position using a so-called "third hand". Or you can use a little screw to fix it to a support. Regardless of the method you choose beware with metal objects not to cause any short circuits.

Next to the serial connector on the printed circuit board of the camera is a jumper - the serial jumper - as shown in the next Figure. Make sure that this jumper is in position.

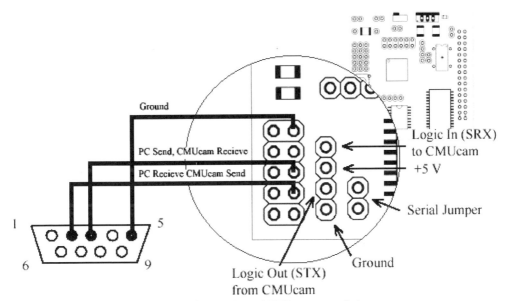

Figure 203. Serial RS232 and TTL ports with jumper.

Start HyperTerm using the following settings, as well as the settings in the next Figure:

Bits per second	115200
Databits	8
Parity	none
Stopbits	1
Datatransportcontrol	none

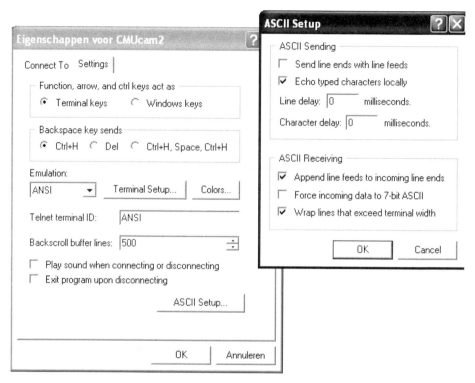

Figure 204. Settings of the HyperTerm connection.

Open the connection and switch the power to the camera on. On the PC the type and version number of your camera appears - see the next Figure. Of course these may be different in your case. If that doesn't happen select a different COM port. Make sure to switch the camera off and back on in between. If that doesn't help connect the RS232 cable on the PCB the other way around, and try the COM ports again.

Enter gv followed by an enter. If what you see is similar to the next Figure the camera is functioning ok, as well as the RS232 connection. When you type it gv will be visible, but the reply from the camera - ACK - will overwrite it.

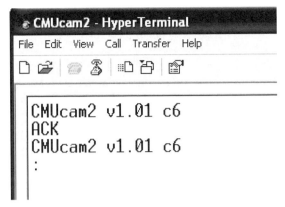

Figure 205. The type number of the camera, and the reply to the gv command.

Close HyperTerm and check to see if Java is installed on your computer. The easiest way to do this is to use the Start menu. Select "execute" and then enter "command" to drop down to DOS, a black window. Now type "Java -version" and press enter. If you get an answer that resembles "Java version" and then a number as you can see in the next Figure than you do have Java. The version number should be 1.4 or higher.

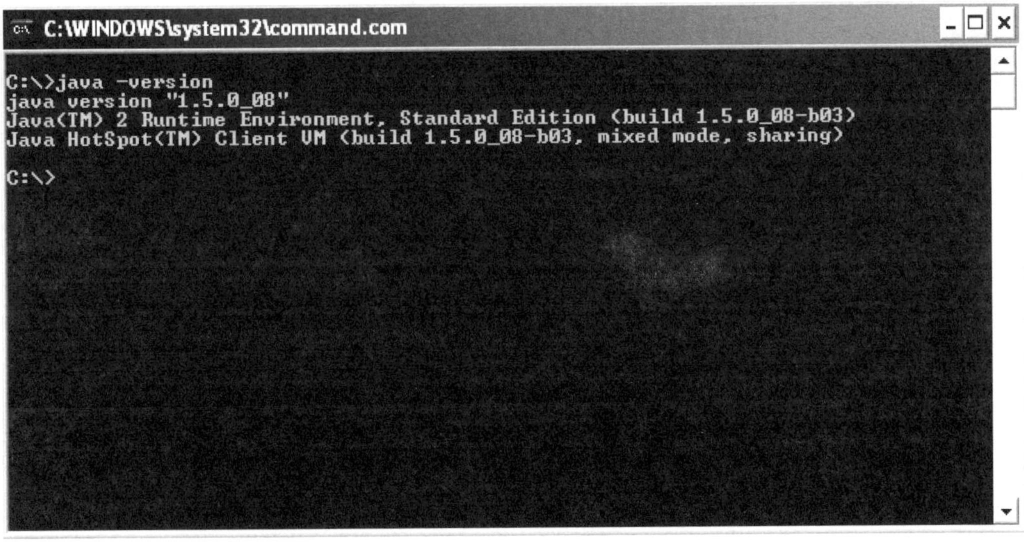

Figure 206. Java version 1.5.0.00 has been installed.

If you get "bad command or filename" or a similar text instead, or your version number is lower than 1.4 you need to go to the website www.java.sun.com to download Java. The name you are looking for is J2SE, JDK or JRE. You will also find instructions as to how Java can be downloaded and installed. Note that for the projects in this book Java is not strictly required, but it is recommended. You can close the black window by typing "exit".

Switch the power to the camera on and go to the CMUcam2 GUI directory and the subdirectory "stand-alone" on your PC. Run the program CMUcan2GUI.jar by double clicking on it.

A small window will pop up. Select the correct COM port and click on OK. Point the camera to something vaguely interesting. In the next window click on Grab Frame.

Figure 207. Grab a frame in the CMUcam2 GUI.

The green LED on the camera starts flashing and after a short while a picture will appear on the screen. In the previous Figure the camera is looking at a small servomotor on my desk. Chances are the picture is slightly out of focus. Turn the little lens slightly left or right to focus the camera and click on Grab Frame again. Keep on doing this until the picture is as sharp as possible. Make sure there is sufficient light. Optionally you can set the camera to high resolution in the "Config" section. This will result in a larger picture that makes focusing easier. The downside is that is takes longer for the picture to be uploaded to the PC. If one of the projects in this book doesn't work as you expect it to work connect the camera to the PC and use this program to see what the camera sees. Usually you can spot right away what the problem is.

This would be a good moment to play with the light to see what lighting setup will give the best visibility of objects in front of the camera. This will come in handy when you do the actual projects. You will achieve the best result when the lights are more or less behind the camera.

Let's see if the camera can see and follow your finger. Select the Motion tab. Click on Load frame and then on Frame Diff. Move your finger in the field of vision of the camera. On the PC screen you will see a green dot in a blue field that is "following" your finger. If this doesn't work properly you most likely need more light, particularly from behind the camera. Once the lights have been changed switch the camera off and back on to try again.

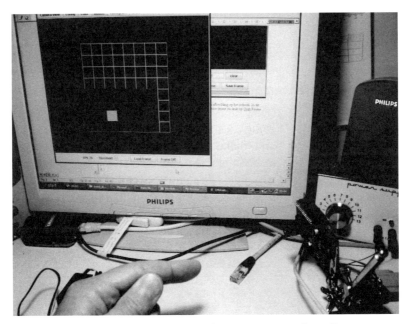

Figure 208. Finger in the lower left corner - seen from the camera

The blue area is the section of the field of vision that the camera is watching, the green dot is what the camera sees as the center of the motion. Click on STOP to stop this test.

If this test is successful too the camera is installed and connected correctly and you can continue with the first project.

Problem solving

You may have run into some problems during these tests:

problem	solution
You cannot get a connection using HyperTerm.	Check the HyperTerm settings, particularly the COM port. Make sure the jumper near the RS232 connector is placed. This jumper and the clock jumper[44] should be the only jumpers placed.
You don't have Java.	Follow the instructions to download Java from the Sun website. Make sure to select a version suitable for your operating system.
The GUI will not respond anymore after clicking Grab Frame.	Press the buttons Ctrl, Alt and Delete on your PC all at the same time. A window pops up. Select applications and then click on CMUcam2GUI. Next click on the button "terminate" to kill the application. Close the window, and try again.
The camera doesn't follow your finger.	Without moving your finger (!) click on Camera View and then click on Grab Frame. Check if the camera can properly see your finger. Use more light, and make sure it shines from behind the camera rather than shining directly into it. Optionally you can use the Config section to enable the white balance. The camera will now correct for poor lighting conditions. You should only do that as a test, make sure there is good lighting when you do the projects. Note that any settings you make will be erased when the power is disconnected.

---

[44] The clock jumper is located next to the Epson chip, consult the CMUcam2 manual.

10.1 Where is my paper?

In this project we will use a CMUcam2 camera to follow an orange piece of paper. Two LEDs will be used to indicate if the paper is to the left or to the right of the center of the field of vision of the camera.

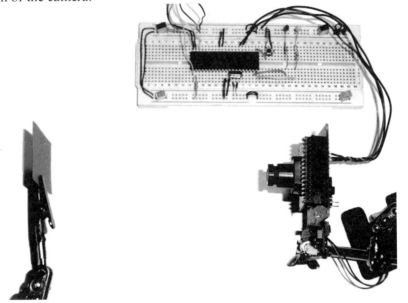

Figure 209. The project with two "third hands".

Technical background

The CMUcam2 camera has a series of image processing commands built in that can be used over a serial connection. You do not need to know for example how the camera is capable of tracking an object with a certain color. You can simply give the TC command (Track Color) and the camera will give you the location.

The easiest way to develop projects with this camera is to connect it to the PC and control it using MICterm. This will allow you to try out your idea step by step, and see how the camera responds. When you are satisfied with the results you can turn your steps into a JAL program. We will use that technique in this book quite often.

When we connect the camera to the microcontroller we will use a speed of 19k2 baud. We will set the camera to that speed before we start our experiments, to make sure that we will see the same phenomena as the microcontroller will see. The baud rate is set by placing two jumpers on the row of contacts next to the power switch, see the next Figure. Start MICterm, select ASCII and switch the camera on.

Figure 210. Two 19k2 jumpers in the circle on the CMUcam2.

By default the camera sends a never ending stream of replies to a single question, because the image that the camera sees may change continuously. The microcontroller must have sufficient time to process these data. So the first thing we will do is make sure we only get one answer. If we want an updated answer we will simply ask the question again. That command is PM 1. Enter that command in MICterm (and press enter). The camera will respond with ACK if the message was understood, or with NCK if it was not understood. It is not unusual for the very first command to be misunderstood, in that case try again. We will make a mental note to ourselves that this may happen, so we will need to address this in our microcontroller program.

You will find an overview of the commands for the camera in the documentation that came with it[45]. The first real step that we need to take is to determine which color the object has. We will assume that the user is holding it straight in front of the camera. Because we don't know how large the object is we will limit the field of vision to the center. For that we can use the command VW. The syntax for this command is VW x1 y1 x2 y2 where x1 and y1 are the left upper corner and x2 and y2 the lower right corner. The next Figure shows the effect of the command VW 35 65 45 75 on the field of vision.

---

[45] If you didn't get any documentation you can download a copy from the manufacturers website: www.cs.cmu.edu/~cmucam/

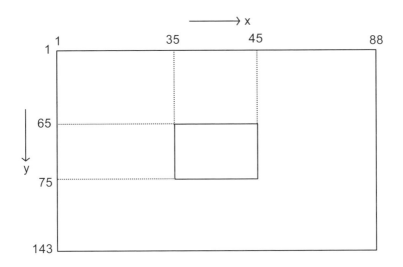

Figure 211. VW 35 65 45 75.

Enter this command in MICterm. If we request the average color using the command GM - Get Mean - then we would get the average color of the small area we just set. If you want you can try the command by entering GM in MICterm. The answer to this command contains not only the three color numbers that we are interested in - red, green and blue - but also much more information, so we will use a mask to suppress everything except for the first three bytes with the colors. The command for masking replies is OM.

OM packet mask \r

This command sets the **O**utput **M**ask for various packets. The first argument sets the type of packet:

#	Tracking Type	Packet
0	Track Color	T
1	Get Mean	S
2	Frame Difference	T
3	Non-tracked packets*	T
4	Additional Count Information**	T, H
5	Track Color Line Mode 2	T
6	Get Mean Line Modes 1 and 2	S

Figure 212. The packet mask.

The first parameter is the packet type, in our case 1 for GM. The second parameter indicates how many bits of information we want to see.

field 8	field 7	field 6	field 5	field 4	field 3	field 2	field 1
128	64	32	16	8	4	2	1

If we choose seven bits than everything is masked except the first three fields (1 + 2 + 4 = 7), and that is exactly what we want. Enter the command OM 1 7 in MICterm, and then request the average color using GM. The reply is ACK followed by an S with three numbers. These three numbers are the quantities of red, green and blue in the color of our piece of paper. So we now know what the color is of the object directly in front of the camera.

We will enlarge the filed of vision of the camera again by issuing the command VW without parameters on MICterm. Before we ask the camera where the color is that we have just measured we will set a new mask. This time all we want is an x and y coordinate of the object with "our" color, so field one and two. The packet number is 1 (for TC) and we want to see the first two fields so the command is OM 0 3.

Depending on the location of the object and the lighting the color may deviate. For that reason we will not ask for an exact color, but for a color range. We will turn the values we received with the GM command into ranges by adding and subtracting 20. If the color value for red was for example 140 we turn this into a range from 140 - 20 to 140 + 20, so 120 to 160. We can now ask where an object is with a color within that range using the command

    TC 120 160 greenlow greenhigh bluelow bluehigh

Where greenlow, greenhigh, bluelow and bluehigh are the rangers of the other two colors.

We only need to do this once, the next time we can just use the command TC for the camera will remember the color ranges. The camera will respond to this command with a T followed by the x and y coordinate of the object with a color within those ranges. In the JAL program we will use the x and y coordinate to light a LED.

This is an overview of the commands that we have used.

command	description
PM 1	For each question just a single reply.
VW 35 65 45 75	Limit the view window to the center.
OM 1 7	Filter GM packets.
GM	Request the average color.
S *red green blue*	Receive the answer.
VW	Maximize the view window.
OM 0 3	Filter TC packets.
TC *r-range g-range b-range*	Convert to ranges by adding +/- 20 to the values.
TC	Request Track Color.
T *horizontal vertical*	Receive the answer.
	And then process it.

The camera replies are meant to be read by humans. For this reason the camera answers with ACK, and numbers are not sent as bytes but as ASCII. When we use the microcontroller to communicate with the camera this will be very inconvenient. Fortunately the camera has a second mode of operation, called "RAW". This has the following effect on the communication:

> *Reply:* Each reply (from the camera) will start with 255 (as a byte), followed by the letter followed by the bytes, all without spaces.
>
> *Command:* Send the two letter command followed by one byte with the number of bytes you will send next, followed by those bytes. This is necessary because for example TC can be used both with variables as well as without. All communication is without spaces.

We select the most extensive RAW mode, including the suppression of ACK. The colon, which serves as a kind of prompt, will still be sent. The command for this mode is RW 7. In RAW mode the previous table looks as follows, where the bold printed numbers are bytes. Space and enter are mentioned explicitly, so the gap between numbers is purely for reading and is not a space. A bold number 35 is ASCII 35 and not a 3 followed by a 5. The ranges - for example r-range - are in reality two bold numbers (for example **0 40**).

command	description
RM *space 7 enter*	Extended RAW mode (ACK also suppressed).
PM **1 1**	For each question just a single reply.
VW **4 35 65 45 75**	Limit the view window to the center.
OM **2 1 7**	Filter GM packets.
GM **0**	Request the average color.
S *red green blue*	Receive the answer.
VW **0**	Maximize the view window.
OM **2 0 3**	Filter TC packets.
TC **6** *r-range g-range b-range*	Convert to ranges by adding +/- 20 to the values.
TC **0**	Request Track Color.
T *horizontal vertical*	Receive the answer.
	And then process it.

You can test this with MICterm as well. The letters can be entered in the send window, the bytes with "Send single byte". This will go so fast that you will not see the indicator turn green. The bytes - which are usually unprintable - appear as question marks in the send window of MICterm.

Hardware

In this project we will use an 18f4455 microcontroller with the camera connected directly to it. We use the built in serial port, so we can keep the passthrough functionality of the Wisp for debugging. The standard serial connection of the 18f4455 is on pins c6 (TX) and c7 (RX). The voltage on these pins is 0 or +5 volts. A normal RS232 connection however uses totally different voltages. The Wisp contains a special chip for the voltage conversion. We could use a chip like that for this application too, but fortunately the camera also has a TTL mode, where TTL basically means 0 or +5 volts. This way a special chip is not needed.

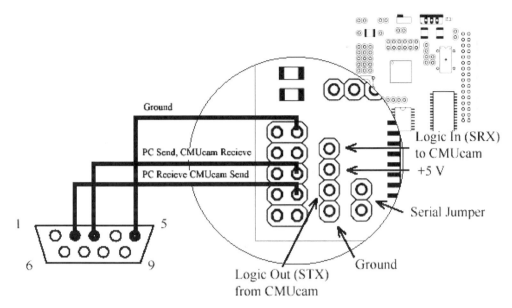

Figure 213. Serial RS232 and TTL gates.

Remove the RS232 cable, and the jumper that is more or less right next to it. The previous and next Figures show the location of that jumper. Next to the RS232 connector you will find a row of four pins. You need to connect the microcontroller to these pins using a small plug. If you do not have such a plug you can also take an 8 pin IC socket and cut it in half. One of these halves will do just fine as a plug (this is how I do it). The connections are as follows:

CMUcam2	18f4455
Logic In (SRX) to CMUcam.	c6 (25)
+5 volts.	Do not use.
Logic Out (STX) from CMUcam.	c7 (26)
Ground.	0 volts on the breadboard.

The +5 volts pin should not be used. This pin could probably be used to power the microcontroller but the voltage regulator on the printed circuit board of the camera is getting rather warm as it is, so I though I'd better use the Wisp. If you use the Wisp you may not connect the +5 pin of the camera to the breadboard as well!

We also need to reduce the baud rate to 19k2 baud, the standard speed of the hardware serial port on the microcontroller as set by the _bert library. This can be done by placing two jumpers on the row near the power switch, see the next Figure. If you did the tests in the technical background section than the speed has already been taken care of.

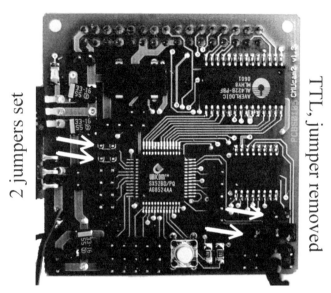

Figure 214. Jumpers and TTL connection.

The rest of the schematic contains no surprises. It is very convenient to use two "third hands" when setting up this project. One for holding the camera and one for holding the piece of paper. Often third hands have a magnifying glass as well as two clips. If you remove the magnifying glass and one clip you are left with a very convenient tool. The picture at the start of this project shows what a suitable setup looks like.

The paper is bright orange, 35 x 35 mm, and approximately 15 cm away from the camera. Of course you can use another color as well as long as it is has a good contrast with the background.

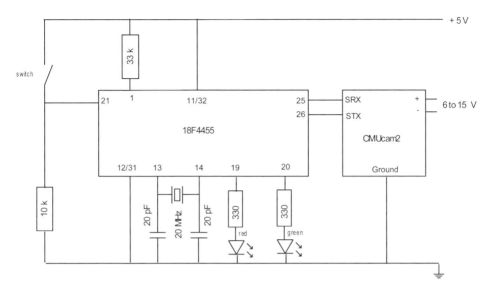

Figure 215. Schematic of the project.

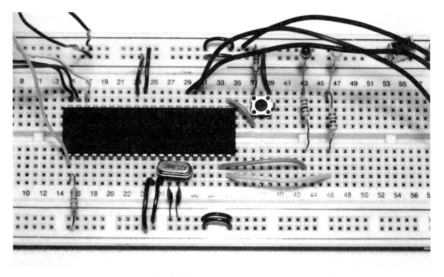

Figure 216. Project on the breadboard.

Software

In this program we will use the hardware serial library that is loaded by the _bert library by default. The content of this library is discussed in section 13.2 of the appendix. The following commands are used in this project:

command	description
Serial_HW_Read(data)	Receive serial data and store it in the variable "data". You can use this command also to check if any data has arrived in the buffer.
Serial_HW_Write(data)	Send the contents of the variable *data*.

While communicating with the camera we will get three types of answers. The colon to indicate that the camera is ready for the next command, a byte with value 255 to indicate that the camera will send a reply and one or more bytes that are the actual reply. For each of these answers a procedure has been made. As an example the procedure that waits for 255.

```
procedure wait_for_packet() is
   var byte data
   data = 0
   while data != 255 loop
      while !serial_hw_read(data) loop end loop
   end loop
end procedure
```

The procedure consists of a while loop inside a while loop. The inner loop will wait for data to come in because as long as there is no data - so !serial_hw_read is true - the program will remain in this loop. This construction is possible because serial_hw_read is a function. This function will look into the RS232 module of the microcontroller to see if any information has come in. If that is not the case then function will return false. Checking the actual content of the variable "data" is therefor pointless because there was no incoming information, so whatever is in variable data it is not going to be something new. If new information has come in however then serial_hw_read will collect the information, store it in the variable "data" and return true. A very smart function, which in the software version unfortunately doesn't work this way. That is because software serial doesn't use a buffer where the function could see if anything came in. So serial_sw_read cannot give true or false

If no information has come in the program will exit the inner loop and arrive in the outer loop. Here we check to see if the information received was a byte with value 255. If that is not the case, so !=255 is true, than the program will stay in the loop and wait for the next information from serial_hw_read. Otherwise it will exit the outer loop and continue with the procedure. Or rather jump back to the main program because the procedure doesn't contain more commands.

The procedure that waits for the colon works the same, the procedure that waits for data is satisfied with any data coming in. These commands are part of the CMUcam2 library.

command	description
wait_for_colon	Wait for the receipt of a colon (:) on the serial hardware connection.
wait_for_packet	Wait for the receipt of a byte with value 255 on the serial hardware connection. This means a packet of data is present.
wait_for_reply(data)	Wait for the receipt of a data on the serial hardware connection and store it in the variable data.

In the technical background section we discovered that the camera often doesn't respond to the first command. So in the program we start with an enter to get its attention.

> serial_hw_write(13)
> wait_for_colon

Then we wait for the colon. Of course the reply contains much more information - such as ACK - but we are not interested in that. When we see the colon we know that the camera is ready for the next command. That command is RM to switch to RAW mode.

> serial_hw_write("R")
> serial_hw_write("M")
> serial_hw_write(" ")
> serial_hw_write("7")
> serial_hw_write(13)
> wait_for_colon

This command must be given with the numbers in ASCII because the camera will switch to RAW mode after processing this command. From now on all commands can be given in RAW mode, the first one being PM 1.

```
serial_hw_write("P")
serial_hw_write("M")
serial_hw_write(1)
serial_hw_write(1)
wait_for_colon
```

As you see the numbers are no longer send as ASCII but as byte. Directly following PM is a one to indicate that one byte will be sent, immediately followed by that one byte, incidentally also a one. Note that we do not need to send an enter (13).

The complete program is in the download packet and too large to print in the book. We will skip a part of the program and move straight to the section where we get data from the camera.

```
-- request average color
serial_hw_write("G")
serial_hw_write("M")
serial_hw_write(0)
wait_for_packet
-- receive the letter, and the three colors
wait_for_reply(reply)
wait_for_reply(red)
wait_for_reply(green)
wait_for_reply(blue)
wait_for_colon
```

We send the GM command without parameters (so GM0) and as a reply we get a letter and three color numbers. The letter ends up in variable reply - but we will not do anything with it. The colors are stored in the correct variables.

The next step is to calculate the ranges. We need to be careful, because all variables are bytes and can only have a value from 0 to 255. If you go over the maximum a byte will simply roll over and start at 0 again. For example 250 + 10 = 5, and so 5 - 10 = 250. So we need to make sure we do not cross these boundaries. If a color number is larger than 20 we can easily deduct 20 to calculate the lower limit of the range. But if it is smaller we must not deduct 20 but simple set the lower limit of the range to zero. The upper limit is similar, except that we use 254 as upper limit because 255 is a reserved value for the camera.

```
if red > 20 then redlow = red - 20 else redlow = 0 end if
if red < 234 then redhigh = red + 20 else redhigh = 254 end if
```

```
if green > 20 then greenlow = green - 20 else greenlow = 0 end if
if green < 234 then greenhigh = green + 20 else greenhigh = 254 end if

if blue > 20 then bluelow = blue - 20 else bluelow = 0 end if
if blue < 234 then bluehigh = blue + 20 else bluehigh = 254 end if
```

Again we skip a part of the program. Once the coordinates posx and posy of the colored piece of paper have been received the proper LED must be lit.

```
if posx > 40 then
    pin_d0 = high
    pin_d1 = low
else
    pin_d0 = low
    pin_d1 = high
end if
```

The rest of the program contains no surprises, and can be found in the download package.

Instructions

1. Hold the piece of paper at a distance of 10 to 15 cm directly in front of the camera.
2. Switch the power on and give the camera time to start up.
3. Press the button on the breadboard.
4. The project is ready for use when both LEDs flash once.
5. Move the paper from left to right. On the left side - as seen from the camera - the green LED will light, on the right side the red LED will light.
6. If it doesn't work the paper wasn't in the center of the field of vision of the camera at step 1.

If you want to reset the project then remember to also reset the camera. Otherwise residual settings may remain that interfere with the operation of the project. Since reprogramming means resetting the microcontroller (but not the camera) make sure to power down the project after programming and then power back up.

Optionally you can send the coordinates to the PC. You must add suitable commands to the program, and use MICterm with the Wisp628/648 option selected. Make sure to start MICterm before pressing the button on the breadboard, because MICterm will reset the microcontroller.

10.2 Count the colored squares

In this project a piece of paper with 1,2 or 3 colored squares will be shown to the camera. The microcontroller will count the number of colored squares and show the answer using three LEDs.

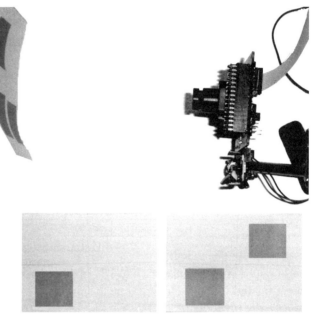

Figure 217. Count the colored squares.

Technical background

Build a test setup using a camera and a rectangular piece of paper. It is probably easiest if you use two third hands, one for the camera and one for the paper. The papers are 10 x 7 cm, and divided in four equally sized sectors. A sector is empty, or has a bright orange square of about 3 x 3 cm on it. There are three different papers with one, two or three orange squares. The distance to the camera is 12 cm.

The next Figure shows how the camera sees such a piece of paper. On the bottom of the screen you can see the color information for the area directly underneath the mouse pointer. This way you can see how the camera perceives the difference between orange and white. Interestingly enough the red value doesn't vary much, but the green value does. On a white section green is about 70, and on an orange section about 30. This may very well have something to do with the lighting in my office - a mains LED lamp.

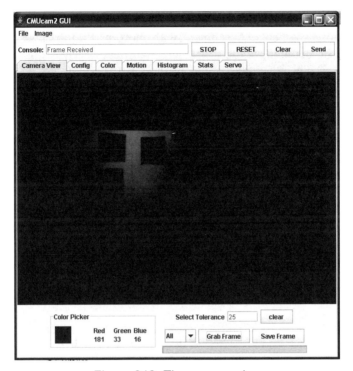

Figure 218. The camera view.

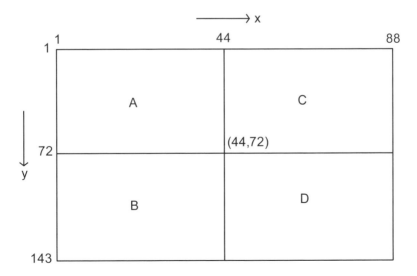

Figure 219. The screen divided in four segments.

In our program we will divide the camera view in four segments, and check the color within each of those segments. A segment with more green than average is apparently orange and must be counted, while a segment with less green is apparently white and must not be counted. The advantage of this method is that you do not need to calibrate for differences in lighting level. The disadvantage however is that papers with none or four orange squares cannot be counted. If you want to be able to count those too you must first determine the color of the orange square, just like in the previous project, and use that information to count. In this project we have chosen a different technique for you already know how to handle the other one.

Just like in the previous project we will make a table of all the commands we need to execute, and test them using MICterm. You do not need any new commands for this project. In this example segment D is the only empty segment. The important answers are shown in the right column: the color information. The first three values are the colors red, green and blue. These are the average values for the entire segment. The next three numbers are the color deviations within that segment, in the order red deviation, green deviation and blue deviation.

command	description
PM 1	For each question just a single reply.
VW 15 15 30 50	Select segment A.
GM	Request average color.
	(S 168 29 16 8 2 0).
VW 15 88 30 120	Select segment B.
GM	Request average color.
	(S 165 28 16 6 2 0).
VW 50 15 70 72	Select segment C.
GM	Request average color.
	(S 175 40 16 7 25 00).
VW 50 80 70 130	Select segment D.
GM	Request average color.
	(S 148 86 16 5 3 0).

In the empty segment D the color green is significantly higher than in the other segments. If the program were to take the average green value based on these measurements - 46 to be exact - then the segments with a green color value lower than that are orange (A, B and C) and the segment with a green value higher than that white (D).

Converted to RAW the table looks like this. As you see the GM mask has been switched on to suppress the deviations.

command	description
enter	Get the camera's attention.
RM *space* 7 *enter*	Extended RAW mode (ACK also suppressed).
PM 1 1	For each question just a single reply.
OM 2 1 7	Mask for the GM packets.
VW 4 15 15 30 50	Select segment A.
GM 0	Request average color.
S *red green blue*	
VW 4 15 88 30 120	Select segment B.
GM 0	Request average color.
S *red green blue*	
VW 4 50 15 70 72	Select segment C.
GM 0	Request average color.
S *red green blue*	
VW 4 50 80 70 130	Select segment D.
GM 0	Request average color.
S *red green blue*	

You can test this with MICterm as well. The letters can be entered in the send window, the bytes with "Send single byte". This will go so fast that you will not see the indicator turn green. The bytes - which are usually unprintable - appear as question marks in the send window of MICterm.

The projects in this book have all been tested, and work. If you design your own projects I highly recommend that you test the final raw table using MICterm, if only just to be sure.

Hardware

In this project we use the 18f4455 microcontroller. Make sure to follow the instructions in the previous project with regard to connecting the camera. Specifically the jumpers for the communication setting - 19k2 - and the TTL connection to the microcontroller.

The camera needs 6 to 15 volts DC, minimum 200 mA capacity. In this project I have used 8 volts. The microcontroller is powered by the Wisp.

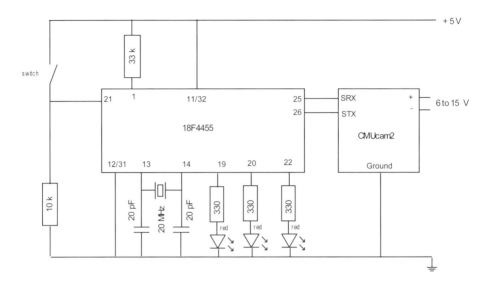

Figure 220. Schematic of the project.

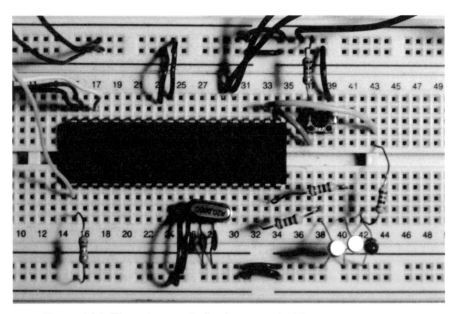

Figure 221. The microcontroller has counted two orange squares.

Software

The software follows the steps in the RAW table. The completed source code is in the download package. Remarkable is the calculation of the average green value:

```
greenAV = 0
greenAV = (greenAV + greenA + greenB + greenC + greenD) / 4
```

First we clear the average value (greenAV), and then we add it to the green values of the individual segments. That seems rather pointless, but the reason for this is that each of the green values is a byte. When solving equations the microcontroller will first solve the right side of the equation. When all variables are bytes it will use a byte as an intermediary value, and then the sum will not fit. It will then divide the result by four, so now it would fit but it is already too late. GreenAV is declared as a word, so by using that on the right hand side the intermediary variable will be a word and now the calculation does fit. So now the sum is correct and we get the correct average.

```
-- count squares
counter = 0
if greenA < greenAV then counter = counter + 1 end if
if greenB < greenAV then counter = counter + 1 end if
if greenC < greenAV then counter = counter + 1 end if
if greenD < greenAV then counter = counter + 1 end if
```

Optionally you can show the average green value as well as the four individual green values on the PC in MICterm. You must start MICterm <u>before</u> you press the button on the breadboard. If you are not interested in that you can delete the following lines from the program. Not that it makes much difference because these are software commands so the microcontroller will not wait for an answer of any kind anyway.

```
serial_sw_write(greenAV)
serial_sw_write(greenA)
serial_sw_write(greenB)
serial_sw_write(greenC)
serial_sw_write(greenD)
serial_sw_write(counter)
```

Finally the answer, the number of orange squares, will be shown using a series of LEDs. When three LEDs are lit the microcontroller counted three orange squares.

```
pin_d0 = low
pin_d1 = low
pin_d3 = low
if counter >= 1 then pin_d0 = high end if
if counter >= 2 then pin_d1 = high end if
if counter == 3 then pin_d3 = high end if
```

Instruction

1. Make sure the paper with the orange squares is properly lit.
2. Switch the power on.
3. Wait a moment to give the camera time to start up.
4. Press the button on the breadboard. The number of orange squares is counted and the correct number of LEDs will be lit.
5. Repeat step 4 as often as you want using the different papers.

Optionally: Start MICterm with settings Wisp6x8 / 1200 baud / add space / raw. MICterm will show the average green value, green A, green B, green C and green D at step 4.

If this project doesn't count correctly adjust the lighting. If that doesn't work execute the commands in the list manually to see if the color differences are detectable in the kind of lighting that you are using.

10.3 I believe something has changed...

Figure 222. Livingroom guard.

In this project we will use a camera to guard a room. It is not a motion detector but a change detector. When something is changed vis-a-vis the original situation the alarm will go off. This will happen also when the change doesn't move, for example an object that is removed from the room.

Technical background

In this project we will introduce two new commands for the camera.

LF \r

See FD on page 34.

This command **L**oads a new **F**rame into the processor's memory to be differenced from. This does not have anything to do with the camera's frame buffer. It simply loads a baseline image for motion differencing and motion tracking.

Figure 223. Load a picture into the camera.

FD threshold \r

See LF on page 39 to load a new baseline frame to difference off of.

This command calls **F**rame **D**ifferencing against the last loaded frame using the **LF** command. It returns a type T packet containing the middle mass, bounding box, pixel count and confidence of any change since the previously loaded frame. It does this by calculating the average color intensity of an 8x8 grid of 64 regions on the image and comparing those plus or minus the user assigned *threshold*. So the larger the threshold, the less sensitive the camera will be towards differences in the image. Usually values between 5 and 20 yield good results. (In high resolution mode a 16x16 grid is used with 256 regions.)

See MD on page 44 to see how to reduce motion noise.

Figure 224. Compare the picture in memory with the real time picture.

The first command is LF, load frame. This command takes as it were a picture of the camera view and stores it in the camera's memory. The second command is FD, frame difference. This command will compare the current view of the camera with the view stored in memory. If anything is different it is compared to a threshold value, and if the change exceeds that threshold a series of measurements is sent to the RS232 connection. The most interesting one of those measurements is the confidence which indicates how certain the camera is that something did indeed change. This value should be larger than zero.

This is not a motion detector command. An object that has been removed, or added for that matter, creates a permanent difference with the image in memory, and thus a permanent alarm.

As in every project we will make a table with a list of commands required to control the camera. Before you continue reading it is perhaps a nice idea when you try to make this list yourself. You can enter it in the next table and test it using MICterm. Use the FD command both with mask as without, and use a threshold of 50.

command	purpose

Then compare your list with mine.

command	purpose
PM 1	For each question just a single reply.
LF	Load a frame in memory (of the camera).
FD 50	Compare the current frame with the one in memory (T 5 5 1 1 8 8 49 45)
OM 2 128	Mask everything but trust
FD 50	Compare the current frame with the one in memory. (T 45)

The threshold value has been arbitrarily set on 50 but in reality this will come from a potmeter. This potmeter can thus be used to adjust the sensitivity of the alarm. In RAW we get the next table:

command	purpose
enter	Get the camera's attention.
RM *space 7 enter*	Extended RAW mode (ACK also suppressed).
PM **1 1**	For each question just a single reply.
LF **0**	Load a frame in memory (of the camera).
OM **2 2 128**	Mask everything but trust
FD **1 50**	Compare the current frame with the one in memory.

Hardware

In this project we use the 18f4455 microcontroller. Make sure to follow the instructions in project 10.1 with regard to connecting the camera. Specifically the jumpers for the communication setting - 19k2 - and the TTL connection to the microcontroller.

The camera needs 6 to 15 volts DC, minimum 200 mA capacity. In this project I have used 8 volts. The microcontroller is powered by the Wisp.

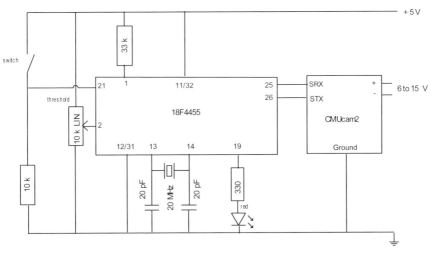

Figure 225. Schematic of the guard.

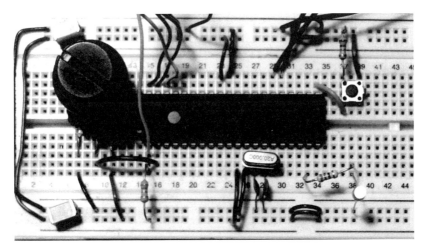

Figure 226. The project on a breadboard.

Software

The interesting part in the code is that the A/D measurement of the potmeter is divided by two to get a suitable range for the sensitivity. This way it is much easier to set the sensitivity just right.

```
-- JAL 2.4j
include 18f4685_bert
include CMUcam2

-- variables
var byte reply, threshold

-- LED
pin_d0_direction = output
pin_d0 = low

-- switch
pin_d2_direction = input

-- wait for the switch to start
while !pin_d2 loop end loop

-- give a dummy enter to get the cameras attention
serial_hw_write(13)
wait_for_colon

-- switch to raw mode but give command in normal mode
serial_hw_write("R")
serial_hw_write("M")
serial_hw_write(" ")
serial_hw_write("7")
serial_hw_write(13)
wait_for_colon

-- one reply per request
serial_hw_write("P")
serial_hw_write("M")
serial_hw_write(1)
serial_hw_write(1)
wait_for_colon
```

```
-- load the first frame
serial_hw_write("L")
serial_hw_write("F")
serial_hw_write(0)
wait_for_colon

-- filter reply packets
serial_hw_write("O")
serial_hw_write("M")
serial_hw_write(2)
serial_hw_write(2)
serial_hw_write(128)
wait_for_colon

forever loop

   -- get threshold
   threshold = ADC_read_low_res(0) / 2

   -- compare actual image with stored one
   serial_hw_write("F")
   serial_hw_write("D")
   serial_hw_write(1)
   serial_hw_write(threshold)
   wait_for_packet
   -- receive the letter, and the confidence
   wait_for_reply(reply)
   wait_for_reply(reply)
   wait_for_colon

   -- show result with LEDs
   if reply > 0 then pin_d0 = high else pin_d0 = low end if

   -- wait a bit
   delay_100ms(1)

end loop
```

Instructions

1. Aim the camera at the object(s) that you want to guard.
2. Start the program.

3. Press on the button on the breadboard.
4. Adjust the sensitivity to the point where the LED is just off. If anything changes in view of the camera the LED will go on. Note that this is not a motion detector. The LED will remain on even if the change is not moving.

10.4 Making pictures for your PC

With this project the microcontroller will take a picture using the camera, and send the picture to the PC. This will happen when you press a button on the breadboard but you could combine it with the previous project and take a picture when an event takes place.

Technical background

We introduce a new camera command: SF. This will send a picture from the camera to the PC.

SF [channel] \r

See FS on page 34, to find out how to stream frames. See DS on page 33, to find out how to reduce data sent by send frame.

This command will Send a Frame out the serial port to a computer. This is the only command that will by default only return a non-visible ASCII character packet. It dumps a type F packet that consists of the raw video data row by row with a frame synchronize byte and a column synchronize byte. (This data can be read and displayed by the CMUcam2GUI java application.) To get the correct aspect ratio, double each column of pixels. Since the image is being read from a buffer, the image resolution is not dependent on baud rate. The baud rate just controls how fast the image will be transmitted. Optionally, a channel (0-2) can be added to the command which causes send frame to only send that channel. This will effectively transmit one third of the data.

Type F data packet format flags:
1 - new frame followed by X size and Y size
2 - new col
3 - end of frame
RGB (CrYCb) ranges from 16-240

| 1 xSize ySize 2 r g b r g b ... r g b r g b 2 r g b r g b ... r g b r g b 3 |

Figure 227. The SF (send a frame) command.

The microcontroller will request this file from the camera when the user presses a button on the breadboard. The command sequence is very simple this time. It is not necessary to switch to RAW mode because the SF command will answer in RAW by default.

command	description
PM 1	For each question just a single reply.
SF	Transfer the picture to the PC.

The microcontroller sends the file directly on to the PC. That means the Wisp programmer is an integral part of the project. Because the camera will communicate with the microcontroller at 19k2 baud the connection between the microcontroller and the PC must be faster, otherwise the camera will "outrun" the PC link and bytes will be lost. For that reason we will change the default speed in the 18f4455_bert library.

1. Copy the 18f4455_bert library from the library directory to the directory where your JAL program is. This is called a "local copy". JAL will first search for libraries in the same directory where the JAL source is, and if that fails it will search in the library directory. By using a local copy only this single program will make use of the modified _bert library. All other programs will still be using the _bert library in the library directory and thus remain fully functional. It is highly recommended to always use this technique if you want to make changes to any library.
2. Search for this command const Serial_SW_Baudrate = 1200 and replace 1200 by 37000 to get this result: const Serial_SW_Baudrate = 37000. This odd baud rate is only applicable to the 18f4455, when you convert this program for example to the 18f4685 or the 16f877A you need to use 38400. Due to the internal clock accelerator in the 18f4455 the baud rate converter in the serial software library doesn't work correctly.
3. Save the modified file.

The picture that was sent to the PC needs to be treated as follows:

    a. Search for a 1.
    b. The next data will be xSize ySize and a 2.
    c. Read red green blue (of a single pixel) until you see another 2 (repeat c) or a 3 (end).
    d. Once the 3 is detected the picture is completed.

The PC program is written in Visual Basic. If you do not own VB5.0 you cannot read the source code but you can of course still use the program.

The Visual Basic communications routine works with events. That means that every time when a (group of) data is received a routine is started. So it is not possible after the receipt of a byte to "request" two more. The PC will have to remember that value one has been detected, which means that the next byte is xSize. And then it must remember that xSize has been detected so now the next byte is ySize. This is a book on microcontrollers and not on PC programming so how this is solved will not be discussed here, but if you are interested you can check out the VB5.0 source code.

Hardware

In this project we use the 18f4455 microcontroller. Make sure to follow the instructions in project 10.1 with regard to connecting the camera. Specifically the jumpers for the communication setting - 19k2 - and the TTL connection to the microcontroller.

The camera needs 6 to 15 volts DC, minimum 200 mA capacity. In this project I have used 8 volts. The microcontroller is powered by the Wisp. The Wisp is an essential part of this project and must remain in place after programming.

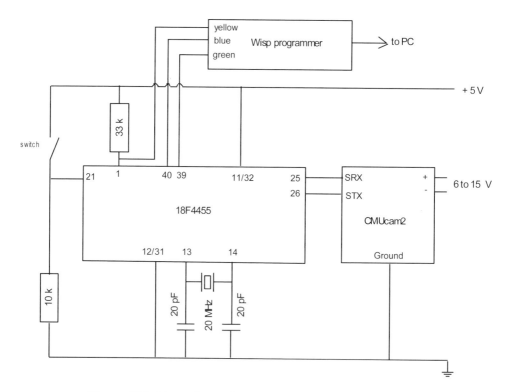

Figure 228. The schematic including the Wisp programmer.

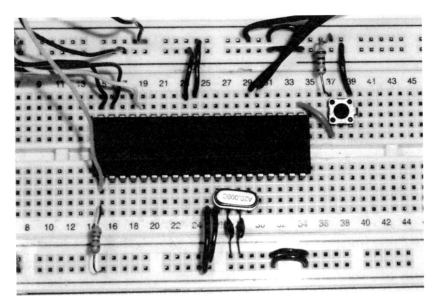

Figure 229. The project on the breadboard.

Software

The JAL program contains no surprises. Once the button on the breadboard is pressed the two commands are sent to the camera. Every incoming byte from the camera is relayed directly to the Wisp and then to the PC. The microcontroller has no idea what it is relaying, so it also doesn't know if the picture is completed yet. Although it would be very easy to find out because the very last byte has value three. The Visual Basic program on the PC is looking for that particular value. As soon as it is detected it will disable the passthrough function of the Wisp, wait a bit and then enable it again. This causes the microcontroller to reset, and thus the program to restart. At that point the program is ready for the next time that the user presses the button.

```
-- JAL 2.4j
include 16f877a_bert
include CMUcam2

-- variables
var byte reply

-- LED
pin_d0_direction = output
pin_d0 = low
```

```
-- switch
pin_d2_direction = input

-- wait for the switch to start
while !pin_d2 loop end loop

-- give a dummy enter to get the cameras attention
serial_hw_write(13)
wait_for_colon

-- one reply per request
serial_hw_write("P")
serial_hw_write("M")
serial_hw_write(" ")
serial_hw_write(1)
serial_hw_write(13)
wait_for_colon

-- request picture
serial_hw_write("S")
serial_hw_write("F")
serial_hw_write(13)
wait_for_colon

forever loop

   while !serial_hw_read(reply) loop end loop
   serial_sw_write(reply)

end loop
```

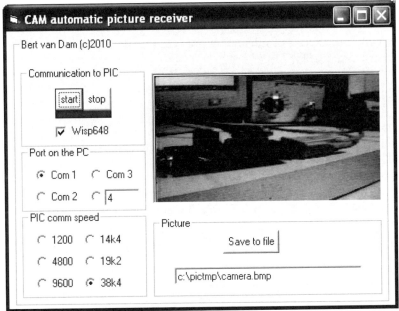

Figure 230. The result on the PC.

Instructions

1. Press the button on the breadboard. Once the picture has arrived on the PC the RS232 connection will be disabled and then re-enabled to reset the microcontroller. This means the button on the microcontroller is operational again.
2. Optionally you can save the picture on the PC by entering a suitable file name (with extension BMP) and pressing on the "Save to file" button.
3. Press the button on the breadboard to take the next picture.

11 Miscellaneous

This chapter contains a series of interesting projects that do not fit in one of the categories that we have discussed so far.

11.1 Seven segment display

This program will count from 0 through to 9 on a single unit of a seven segment display.

Technical background

This display consists of seven LEDs in the shape of a line called segments, hence the name seven segment display. The segments are placed in such a way that they can be used to make numbers and letters, though for the latter it does require a bit of fantasy.

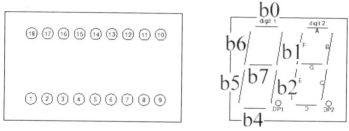

The digits are common anode (+). The pinout is shown below (top view!).

digit 1 segment E	1	18	digit 1 segment F
digit 1 segment D	2	17	digit 1 segment G
digit 1 segment C	3	16	digit 1 segment A
DP1	4	15	digit 1 segment B
digit 2 segment E	5	14	digit 1 common (+)
digit 2 segment D	6	13	digit 2 common (+)
digit 2 segment G	7	12	digit 2 segment F
digit 2 segment C	8	11	digit 2 segment A
DP2	9	10	digit 2 segment B

Figure 231. Pin layout of the 7-segment display.

For one of the displays the link between segment and microcontroller pin in shown in the previous Figure. Pin b3 is skipped because this is the LVP pin that some programmers use.

The Figure - from the datasheet - also shows that all segments have a common anode (+) which means that one side of the segments is connected to the plus. That is inconvenient because it means that you need to make a pin low to light the segment. That is the only way to connect a LED to + and ground at the same time.

In principle it is now very simple. For every digit we need to set specific pins high and others low. There are 10 possibilities - 0 trough to 9 - so we could use ten "if.. then..." statements. Fortunately there is an easier way: the "case" statement. With this statement you can use a single condition - in our case a variable - to take a whole range of decisions.

```
case number of
   1: block
      pin_b0 = 1
      pin_b1 = 0
      pin_b2 = 0
      pin_b4 = 1
      pin_b5 = 1
      pin_b6 = 1
      pin_b7 = 1
   end block
   2: block
      pin_b0 = 0
      pin_b1 = 0
      pin_b2 = 1
      pin_b4 = 0
      pin_b5 = 0
      pin_b6 = 1
      pin_b7 = 0
   end block
end case
```

In this code one of the two blocks is selected based on the value of variable "number". If variable number has value one then the first block will be executed, if variable number has value two then the second block will be executed. If number has an other value then none of the blocks will be executed. If you take a closer look at case one you will note that the segments that are switched on form the number one (remember: pins that are low make a segment go on). Using this technique you can make a case statement for each number, and if you are bored for each letter too, and put it in a library. This can be done manually but it is much easier to use a program from the download package called "7segment text maker".

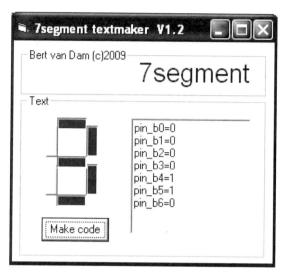

Figure 232. 7-segment text maker.

Using the mouse you can click on segments to turn them on or back off again. If you like what you see click on the "Make code" button, and the little window shows that code that you need to use to make this text. The program "knows" that zero is on and one is off, so it produces the correct code for this situation. Using the mouse you can copy the code from the little window into your JAL program or library.

Hardware

So far we usually connected just a few LEDs to the microcontroller, but this time we connect seven at the same time so we need to take a closer look at the total power consumption.

Using a multimeter we can determine that a single segment uses 13 mA of current when a current limiting resistor of 220 ohm is used. That means that the forward voltage over a single segment is 2.9 volts, much more than a normal LED. The segment is still visible with a current limiting resistor of 680 ohm, but not very well. An ideal value where the segments are clearly visible and the current is still low seems to be 470 ohm. Using that resistance the current per segment is 6 mA, way below the maximum of 25 mA per pin that the microcontroller can handle. For eight segments (seven plus the dot in case you want to use that too) the current is 49 mA still well below the maximum of 200 mA, the maximum the microcontroller can source.

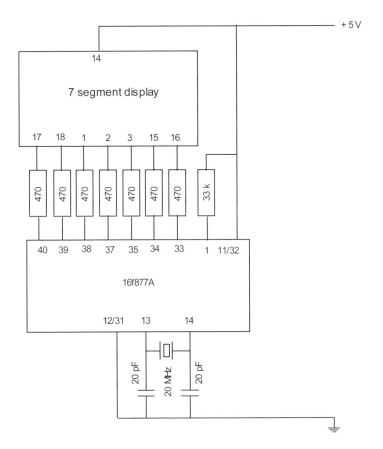

Figure 233. Schematic of the 7-segment display on a 16f877A microcontroller.

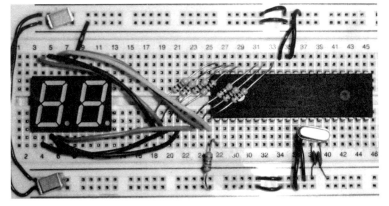

Figure 234. Project on the breadboard, only the left digit is connected.

Software

The 7segment library contains all digits from 0 through to 9.

command	description
segnumber(n)	Show the number n on a 7-segment display. See the library for the connections of the segments to the pins.

This library makes our program very simple. A counter shows the numbers one by one for one second on the left display.

```
-- JAL 2.4i
include 16f877A_bert
include 7segment

-- define the variables
var byte counter

forever loop

  -- show all numbers
  for 10 using counter loop
    segnumber(counter)
    delay_1s(1)
  end loop

end loop
```

11.2 Two 7-segment display's with transistor switching

This program measures the position of a potmeter connected to pin a0 and displays this in two digits on the 7-segment display.

Technical background

In this project we will use both sides of the 7 segment display. We could of course connect the second unit to another port of the microcontroller, the 16f877A has enough pins. But what if we want to make a clock with four digits, or maybe even six. This would result in a huge amount of pins and power consumption!

Fortunately there is a simpler way: we connect both units to the same pins. The joint connector for each unit is not connected to +5 volts but to a transistor. These two transistors decide which part of the display is on and which is off. If we switch between the two parts of the displays fast enough it looks like both parts are always on.

The first step is to decide which transistor we need, using the steps in section 13.6 of the appendix.

1a. The current per segment is 6 mA, and seven are in use (the dot is not connected) which means a total of 42 mA. So the transistor needs to have an $I_{c(max)}$ of at least 42 mA.

1b. The total current is 42 mA at 5 volts, so the resistance of the unit is 5 V / 42 mA = 119 ohm, which is R_L

2. The maximum current that the microcontroller can deliver is on a single pin 25 mA, so the transistor needs to have an amplification of at least $h_{FE}(min)$ = 5 * (42 mA / 25 mA) = 8.4

3. In the appendix a small list of transistors is shown. We select the BC547B with an $I_{C(max)}$ of 500 mA and an $h_{FE(min)}$ of 200, and because I only have one of those we also select the BC547C which is identical but has an $h_{FE(min)}$ of 420.

4. Because both microcontroller and transistor are connected to the same V_{DD} (+ 5 volts) we can use the simplified formula for the base resistor.

$R_B = 0.2 \times R_L \times h_{FE}$ so $R_B = 0.2 \times 119 \times 200 = 4k7$
for the B version and 10k for the C version.

For the C version using a multimeter the measured current through the base resistor is 0.02 to 0.05 mA depending on the number of segments that is on. That matches our theory nicely: a collector current of 43 mA if all segments are on and an amplification of 420 means a base current of 43/420 = 0.1 mA. This current is on only 50% of the time (the rest of the time the other unit is on), and 50% of 0.1 = 0.05 mA.

Hardware

In this project we will use a 16f877A microcontroller.

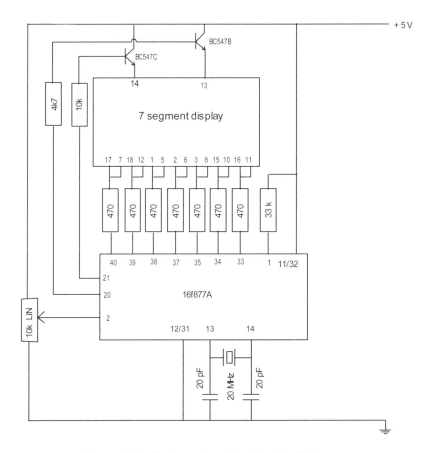

Figure 235. Schematic of the double digit setup.

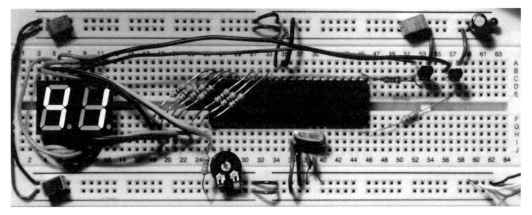

Figure 236. The number 41 on the 7-segment display.

Software

To make the digits look good they must be on long enough to see them well, and short enough to prevent flashing. A good value turns out to be 10 mS. So the program will show one digit for 10 mS, then the other digit, then the first one again et cetera for a period of 0.1 seconds. Feel free to experiment with then timing if you want to see if you like other values better.

The other interesting thing in the program is the conversion of the A/D measurement from a range of 0 - 255 to a range of 0 - 99, because we only have two digits available. This is a bit more complicated than it seems.

    -- convert to 99 range
    value = (word(resist) * 39) / 100

On a calculator you could simply multiply the A/D measurement by 99/255 = 0.39. In a microcontroller that is impossible because we cannot use decimals. The solution is to turn the decimal value 0.39 to a faction 39/100. Now we can multiply by 39 and then divide by 100 and get the correct result. Of course this will only work well if you use words rather than bytes. In previous projects we have seen that we can accomplish this by using a variable defined as a word on the right hand side of the equation. Since we do not have such a variable at hand we will use a different technique. We can force the compiler to treat the byte variable "resist" as a word by using this syntax: word(resist). So now the calculation will fit.

    -- split over the two units
    counter1 = value/10
    counter2 = value-(counter1 * 10)

When splitting a number in tens and ones we make use of the fact that bytes can only be whole numbers. Let's say the variable "value" is 47. That means for counter1 and counter 2:
    counter1 = 47/10 = 4
    counter2 = 47 - (4 * 10) = 7

This is the completed program:

    -- JAL 2.4i
    include 16f877A_bert
    include 7segment

```
-- pins for transistor control
pin_d1_direction = output
pin_d2_direction = output

-- turn transistor control off
pin_d1 = 0
pin_d2 = 0

-- define the variables
var byte counter1,counter2, resist
var word value

forever loop

   -- convert analog on a1 to digital
   resist = ADC_read_low_res(0)

   -- convert to 99 range
   value = (word(resist) * 39) / 100

   -- split over the two units
   counter1 = value/10
   counter2 = value-(counter1 * 10)

   -- show each digit 1 second
   -- nl 100 x 10 mS
   for 10 loop

     -- left digit
     pin_d2 = 1
     segnumber(counter1)
     delay_10mS(1)
     pin_d2 = 0

     -- right digit
     pin_d1 = 1
     segnumber(counter2)
     delay_10mS(1)
     pin_d1 = 0

   end loop
end loop
```

11.3 Rotary Encoder

In this project we will read a rotary encoder and use an RC network to eliminate bouncing. In the next project we will use a port B interrupt in combination with a rotary encoder.

Technical background

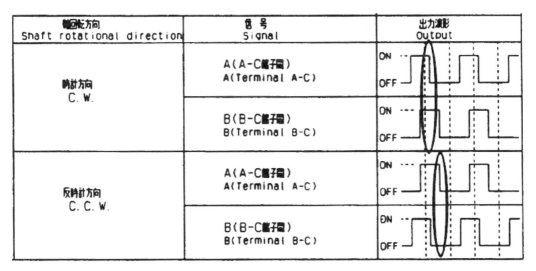

Figure 237. Rotation schedule from the datasheet of the rotary encoder.

The previous Figure is from the datasheet of the Alps Electronic C. rotary encoder. In this program we will wait for the negative edge of contact A - the upper square wave in each picture. We will find this edge by waiting until the level is high, and then waiting until it is low again.

> while !pin_b0 loop end loop -- wait for a low-high edge
> while pin_b0 loop end loop   -- wait for a high-low edge

You might perhaps think that does would do the trick as well:

> while pin_b0 loop end loop

But this is also true if pin b0 went down "some time ago" rather than just yet. Since we have no control over the exact moment that we start looking we need to make sure that we have the actual edge, and not just a low level.

When the negative edge occurs we look at contact B. If that contact is high the rotary encoder is rotating CW - Clockwise - as you can see in the upper pictures. If contact B is low the rotary encoder is rotating CCW - Counter Clockwise - as you can see in the lower pictures.

In theory this will work perfectly. In practice however bouncing of the switch has a devastating effect. Bouncing means that the contacts inside the rotary encoder aren't closing properly but momentarily open and close several times before ending in a stable position. This results in a whole series of pulses with cause erroneous results.

The solution we have opted for is to use an RC network consisting of a resistor of 47 k (yellow - purple - orange) as pull up resistor and a 220 nF capacitor over the microcontroller pins to the ground, as close as possible to the microcontroller. The next Figure shows this network for one switch of the rotary encoder[46].

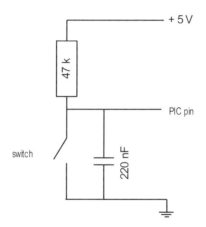

Figure 238. Debouncing network.

The RC network works as follows. As long as the switch is open the capacitor is charged. When the contact closes the capacitor immediately discharges through the switch. If the switch opens again the charging of the capacitor takes time because the current has to go through the resistor. That means it takes a moment before the pin is actually high. Rapid opening and closing of the contact has therefor no effect on the pin because it will stay low until the switch is open long enough for the capacitor to charge. The network we use has the following time constant:

$$t = R \times C = 47k \times 220 \text{ nF} = 47 \cdot 10^3 \times 220 \cdot 10^{-9} = 0.01 \text{ sec}$$

---

[46] Of course you can use this same technique to debounce normal switches.

The manufacturer claims in the datasheet that the bouncing takes a maximum of 2 mS. If we assume that during that time the switch bounces two times, then the bouncing time itself is 1 mS. The voltage over the charging capacitor can be calculated using the time constant (0.01 second or 10 mS) using this formula:

$$V_t = v (1 - e^{(-t/RC)})$$

When t is equal to 1 mS and the voltage is 5 volts then this formula becomes:

$$V_t = 5 (1 - e^{(-1/10)}) = 0.48 \text{ volts.}$$

The microcontroller will regard voltages of over 2 volts as high, so we are sufficiently far from that. According to the datasheet the microcontroller will see a pin as low when the voltage is below $0.15 * V_{DD}$ for a normal I/O pin in a 16f877A. When V_{DD} is 5 volts that results in 0.75 volts. The area in between (0.75 to 2 volts) is undefined, also know as the forbidden range. We are far below that value as well. In the - rather likely - event that the switch bounces more often the voltage will be even lower than 0.48 volts. The network is therefor very effective in debouncing this switch.

In principle a switch can be damaged if the current passing through it is too large. Discharging a capacitor can cause such a large current. Even when rotating the rotary encoder very fast a current of just 0.2 mA is measured over the contacts of the switch and that seems low enough not to cause damage.

Hardware

In this project we will use the 16f877A microcontroller.

When you are building this project on a breadboard, or a printed circuit board, it is very important that the capacitors of the RC network are placed as close as possible to the pins of the microcontroller. The resistors are best placed near the rotary encoder. Just as you can see in the next Figure.

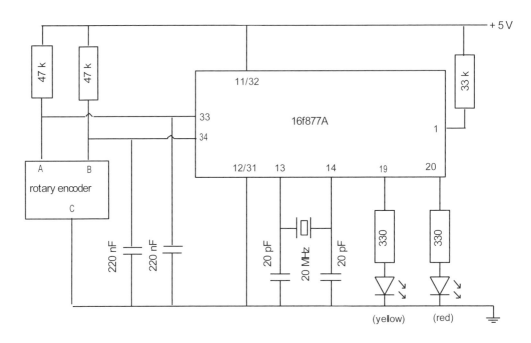

Figure 239. Schematic of the rotary encoder.

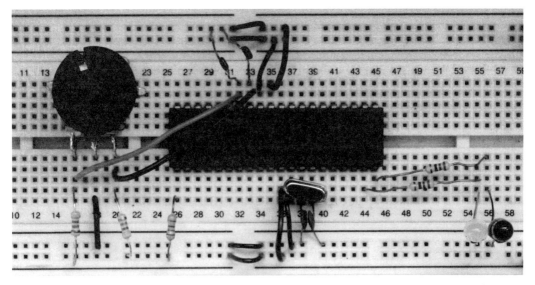

Figure 240. The rotary encoder on a breadboard, not the positions of R and C.

Software

The direction of rotation of the rotary encoder is shown using two LEDs. Red means that the rotary encoder is turned to the right - or clockwise - yellow means that it is turned to the left - or counter clockwise. The result can also be displayed on the PC in MICterm as an incremental counter. That means the counter will increase when you turn to the right and decrease when you turn to the left. To visualize this signal use "RAW" as setting in MICterm.

```
-- JAL 2.4i
include 16f877A_bert

pin_b0_direction = input      -- rotary terminal AC
pin_b1_direction = input      -- rotary terminal BC
pin_d0_direction = output     -- yellow LED (Clockwise)
pin_d1_direction = output     -- red LED (Counter Clockwise)

var bit cw is pin_d0          -- uses names from the datasheet
var bit ccw is pin_d1

var byte moved                -- movement counter
moved = 125                   -- start in the middle

forever loop

    while !pin_b0 loop end loop  -- wait for a low-high edge
    while pin_b0 loop end loop   -- wait for a high-low edge

    if pin_b1 == 1 then
        cw = high                -- indicate cw move
        ccw = low
        moved = moved + 1        -- add one to movement counter
    else
        cw = low
        ccw = high               -- indicate ccw move
        moved = moved - 1        -- substract one from movement counter
    end if

    serial_sw_write(moved)       -- send data to PC
    delay_10ms(25)               -- wait a bit

end loop
```

11.4 Port B interrupt

In several projects in this book we discussed the timer interrupt. There is another important interrupt and that is the port B interrupt. In this project we will use a port B interrupt on a 16f877A to read a rotary encoder. .

Technical background

When a program is for example executing a delay command signals on input pins may pass unnoticed. Some of the pins on port B can be connected to an interrupt routine. When a change on such a pin occurs and interrupt procedure is executed.

In order to use a port B interrupt we first need to set two registers, just like with the timer0 interrupt, namely option_reg and intcon.

OPTION_REG REGISTER

R/W-1	R/W-1	R/W-1	R/W-1	R/W-1	R/W-1	R/W-1	R/W-1
$\overline{\text{RBPU}}$	INTEDG	T0CS	T0SE	PSA	PS2	PS1	PS0
bit 7							bit 0

bit 7	**$\overline{\text{RBPU}}$**
bit 6	**INTEDG**
bit 5	**T0CS**: TMR0 Clock Source Select bit 1 = Transition on T0CKI pin 0 = Internal instruction cycle clock (CLKOUT)
bit 4	**T0SE**: TMR0 Source Edge Select bit 1 = Increment on high-to-low transition on T0CKI pin 0 = Increment on low-to-high transition on T0CKI pin
bit 3	**PSA**: Prescaler Assignment bit 1 = Prescaler is assigned to the WDT 0 = Prescaler is assigned to the Timer0 module
bit 2-0	**PS2:PS0**: Prescaler Rate Select bits

Bit Value	TMR0 Rate	WDT Rate
000	1 : 2	1 : 1
001	1 : 4	1 : 2
010	1 : 8	1 : 4
011	1 : 16	1 : 8
100	1 : 32	1 : 16
101	1 : 64	1 : 32
110	1 : 128	1 : 64
111	1 : 256	1 : 128

Figure 241. The option_reg register of the 16f877A.

In this register the highest bit - internal pull-up resistor - is important. We definitely do not want an internal pull-up resistor because the pin is connected to a RC network. Oddly enough this bit must be set to one if you do not want pull-up resistors. Bit 6 can be used to determine if the interrupt on pin b0 - which we will enable in a minute - should occur on the rising or on the falling edge. In the previous project we used the falling edge, so we will use it here too. That means this bit must be zero. The remaining bits are for timer0 so we can clear those too.

option_reg = 0b_1000_0000

INTCON REGISTER (ADDRESS 0Bh, 8Bh, 10Bh, 18Bh)

R/W-0	R/W-0	R/W-0	R/W-0	R/W-0	R/W-0	R/W-0	R/W-x
GIE	PEIE	T0IE	INTE	RBIE	T0IF	INTF	RBIF
bit 7							bit 0

bit 7 **GIE:** Global Interrupt Enable bit
    1 = Enables all unmasked interrupts
    0 = Disables all interrupts

bit 6 **PEIE**: Peripheral Interrupt Enable bit
    1 = Enables all unmasked peripheral interrupts
    0 = Disables all peripheral interrupts

bit 5 **T0IE**: TMR0 Overflow Interrupt Enable bit
    1 = Enables the TMR0 interrupt
    0 = Disables the TMR0 interrupt

bit 4 **INTE**: RB0/INT External Interrupt Enable bit
    1 = Enables the RB0/INT external interrupt
    0 = Disables the RB0/INT external interrupt

bit 3 **RBIE**: RB Port Change Interrupt Enable bit
    1 = Enables the RB port change interrupt
    0 = Disables the RB port change interrupt

bit 2 **T0IF**: TMR0 Overflow Interrupt Flag bit
    1 = TMR0 register has overflowed (must be cleared in software)
    0 = TMR0 register did not overflow

bit 1 **INTF**: RB0/INT External Interrupt Flag bit
    1 = The RB0/INT external interrupt occurred (must be cleared in software)
    0 = The RB0/INT external interrupt did not occur

bit 0 **RBIF**: RB Port Change Interrupt Flag bit
    1 = At least one of the RB7:RB4 pins changed state; a mismatch condition will continue to set the bit. Reading PORTB will end the mismatch condition and allow the bit to be cleared (must be cleared in software).
    0 = None of the RB7:RB4 pins have changed state

Figure 242. The intcon register of the 16f877A.

In the intcon register interrupts in general must be switched on using bit 7. We are particularly interested in pin b0 because that is what contact AC is connected to, so we will set bit 4 too.

        intcon = 0b_1001_0000

When the interrupt routine is called we need to make sure that this is caused by a port B interrupt. For this we use flag intf. It is also important to check if pin b0 is indeed zero.

So the basic interrupt structure looks like this:

```
procedure detect_a_pin is    -- the actual interrupt routine
   pragma interrupt
                             -- check if this interrupt is b0 (intf)
                             -- and is on a high-low transition (pin_b0 == 0)
   if (intf) & (pin_b0 == 0) then

      [ the commands ]

      intf = 0               -- clear intf to re-enable pin b0 interrupts

   end if
end procedure
```

At "the commands" we will put the code that keeps track of the rotation direction and controlling the LEDs. Normally speaking an interrupt routine should be kept as short as possible, so this is not very elegant. But it is easy to read and understand. And because we use variable flag to prevent multiple entries the length of the interrupt routine is technically irrelevant. So we will opt for functionality over beauty.

Hardware

The hardware is identical to the previous project, including the debouncing networks.

Software

The direction of rotation of the rotary encoder is shown using two LEDs. Red means that the rotary encoder is turned to the right - or clockwise - yellow means that it is turned to the left - or counter clockwise. The result can also be displayed on the PC in MICterm as an incremental counter. That means the counter will increase when you turn to the right and decrease when you turn to the left. To visualize this signal use "RAW" as setting in MICterm.

```
-- JAL 2.4i
include 16f877A_bert

pin_b0_direction = input    -- rotary terminal AC
pin_b1_direction = input    -- rotary terminal BC
pin_d0_direction = output   -- yellow LED (Clockwise)
pin_d1_direction = output   -- red LED (Counter Clockwise)

var bit cw is pin_d0        -- uses names from the datasheet
var bit ccw is pin_d1

var byte moved,lastmoved    -- movement counter

moved = 125                 -- start in the middle

procedure detect_a_pin is   -- the actual interrupt routine
   pragma interrupt

                            -- check if this interrupt is is b0 (intf)
                            -- and is on a high-low transition (pin_b0 == 0)
   if (intf) & (pin_b0 == 0) then

      if pin_b1 == 1  then
         cw = high          -- indicate cw move
         ccw = low
         moved = moved + 1  -- add one to movement counter
      else
         if pin_b1 == 0  then
         cw = low
         ccw = high         -- indicate ccw move
         moved = moved - 1  -- subtract one from movement counter
         end if
      end if

      intf = 0              -- clear intf to re-enable pin b0 interrupts

   end if
end procedure
```

```
   option_reg = 0b_1000_0000   -- disable pull-up and high-low edge detect

   intcon = 0b_1001_0000       -- enable interrupts general and port B

   forever loop

      if moved != lastmoved then  -- did the data change since last time

         lastmoved = moved           -- remember current value of moved
         serial_sw_write(moved)      -- send data to PC

      end if

   end loop
```

11.5 Upgrade your Wisp programmer firmware

In this project we will show you how to upgrade the software - also known as "firmware" - in your Wisp programmer.

Technical background

Figure 243. The Wisp648 programmer, the chip on the left is the 16f648.

The Wisp is an intelligent programmer. This intelligence is based on software that runs in the programmer itself. The Wisp648 uses a 16f648A microcontroller. The software in the programmer is called firmware.

New firmware versions appear on a regular basis. You can find the official versions on the website of the manufacturer www.voti.nl. On the internet you may also find unofficial versions. You are free to use any version you like, but you are probably better off using the official versions. That way you can be certain that software and hardware match perfectly.

The firmware in the programmer works together with software on the PC called Xwisp, which is also regularly updated with new versions. New versions often support more microcontrollers or provide extra functionality. Both the Wisp firmware as well as Xwisp upgrades are free.

Please note: if you own the Wisp628 programmer you can still make use of the latest firmware, by replacing the 16f628 in your programmer by a new 16f648A. That means you follow the instructions in this chapter and at the end take the 16f628 out of your programmer and replace it by the 16f648A that you just programmed.

Hardware

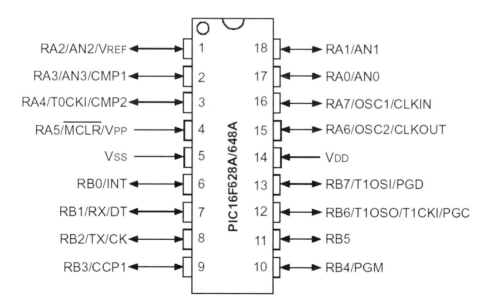

Figure 244. Pin layout of the 16f648A microcontroller.

Build a basic setup for the 16f648A and connect the programmer as follows:

pin	color wire
4	yellow
10	white
12	green
13	blue

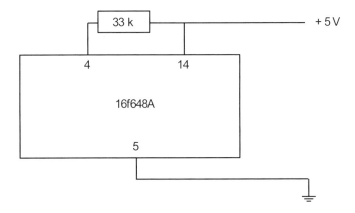

Figure 245. Upgrade hardware.

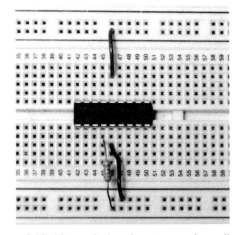

Figure 246. Upgrade hardware on a breadboard.

Instructions

1. Download the latest version of the firmware from the website of the manufacturer www.voti.nl. Normally speaking this file should be called wisp648.hex. In the download package you will find the firmware that was used for all programs in this book in the directory of this section: 11.5.
2. Build the basic setup as shown in the hardware section.
3. Connect the Wisp programmer.
4. Start the program xwisp_gui.exe in the directory c:\picdev2\xwisp. The black window in the background is normal, it will close automatically when you are done with the program.
5. Select "File" and then "Load" and select the file you downloaded or the file from directory 11.5 wisp648.hex (make sure it has the correct extension).
6. Click on the "Go" button and wait for the text "OK" to appear. This may take a while.
7. Switch off the power and remove the plug from the Wisp programmer.
8. Remove the 16f648A from the programmer and replace it with the 16f648A that you just programmed. Note the location of the notch, it should be near the edge of the Wisp printed circuit board.[47]
9. Keep the old 16f648A so you can use it again for the next upgrade.

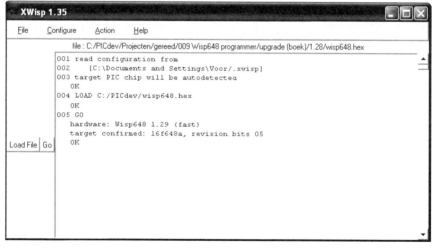

Figure 247. Upgrade using the xwisp_gui.

---

[47] If you have the Wisp628 you need to remove the 16f628 in this step. The new firmware doesn't fit in this microcontroller. If you keep it you can temporarily put it back in the Wisp programmer next time you want to upgrade the 16f648A again.

11.6 Laser alarm

A laser is a convenient device to use in an alarm. The laserbeam stays small and focussed, so you can make a long route through the area you want to protect using mirrors. This way large area's can be effectively guarded.

Technical background

In this project we use a little surplus laser of about 5 by 10 mm with a red laserbeam with a diameter of about 4 mm. The module is designed for 6 mA and has a forward voltage of 2.5 volts. That means the current limiting resistor should be:

$R = V/I = (5-2.5)/0.006 = 417$ ohm.

We select a slightly larger value of 470 ohm. The current is low enough to connect the laser directly to the microcontroller, but that serves no purpose in this project so we will connect it directly to the power supply.

The laserbeam is, after a long route with some mirrors, collected with a light dependent resistor, or LDR. The relationship between the amount of light and the resistance is usually something like this:

R (in k ohm) $= 500 / L$ (in Lux)

To get a feeling what Lux means a small table with some examples.

location	light intensity
direct sunlight	100.000 Lux
overcast day	10.000 Lux
office	350 Lux
room with a candle	50 Lux

Table 48. Some examples of Lux in different places.

Even "identical" LDRs can differ as much as 50% so you always need some sort of calibration routine or potmeter. Good to remember: more light means lower resistance.

A power dissipation of 50 tot 80 mW is quite normal for an LDR. At 5 volts this translates to 10 to 16 mA, so the LDR can be connected directly to the pin of a microcontroller.

Hardware

In this project we will use a 16f877A microcontroller. The LDR is connected to the ground and at the other side to pin 2 of the microcontroller, so A/D converter 0. Since an A/D converter can only measure voltages and not resistance the LDR needs to be connected to the +5 volts through a resistor. The microcontroller is now connected to the middle of a voltage divider. Depending on the resistance of the LDR the voltage on the center of the voltage divider will vary. Instead of a fixed resistor a potmeter is used, so we can use the potmeter also for calibration.

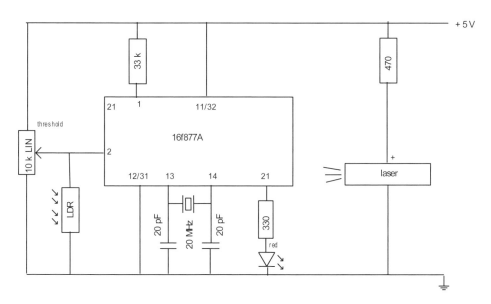

Figure 248. Schematic of the laser alarm.

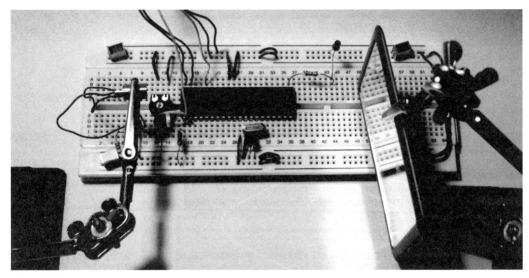

Figure 249. The project with the mirror real close (to get it in the picture).

In the previous Figure both the mirror and the laser are held in position using a third hand. The laserbeam is reflected by the mirror onto the LDR as shown in the next Figure.

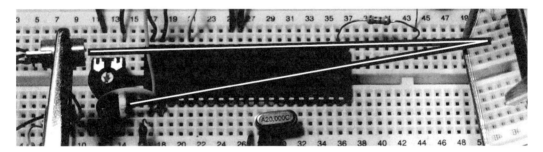

Figure 250. Trajectory of the laserbeam from laser to LDR.

In practice the mirror is of course much further away, and firmly mounted on for example a wall, just as the LDR and the laser itself. It is best to adjust the laserbeam trajectory using the mirror in the dark. The laserbeam will be a bright red spot that should be centered on the LDR. Adjust the potmeter so that the LED is just off. Now if you intercept the laserbeam for example with your hand the LED will go on.

> Be very careful: laserbeams can damage your eyes, even from a large distance. Be careful when adjusting the mirror(s) and place laser, mirror(s) and LDR in such a way that nobody can look into the beam, including animals. Use a small and light mirror and carefully tighten the screws in the third hand before testing this project.

Software

The software is very simple: light the LED when a threshold value on the LDR is exceeded.

```
-- JAL 2.4j
include 16f877a_bert

-- definitions
pin_a0_direction = input
pin_d2_direction = output

--define variables
var byte laser

forever loop

   -- get laser reading
   laser = adc_read_low_res(0)

   -- path interrupted?
   if laser > 90 then
      pin_d2 = high
   else
      pin_d2 = low
   end if

end loop
```

12 Other microcontrollers

In this book the 16f877A, 18f4455 and 18f4685 microcontrollers are used, but of course there are many more types available. For about 15 microcontrollers _bert libraries have been made. One of the advantages of this library is that it is relatively easy to change from one microcontroller to another.

12.1 Supported microcontrollers

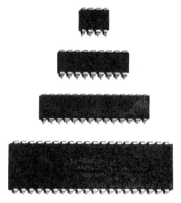

Figure 251. Top down: 10f200, 16f628, 16f876A, 18f4685.

Data

The next table applies only when the _bert library for that particular microcontroller is loaded. Because of this every microcontroller has software RS232 capabilities- except the 10f200 which is too small. So if the table shows a 1 in the RS232 column this refers to the software option. If the table shows 2 (or more) then this microcontroller also has a hardware RS232 unit.

Note that the I^2C library and SPI library from the appendix can also be used on all microcontrollers - again except the 10f200. So if the table shows a 1 in these columns this refers to the software options - note that you must load these libraries to get these options. If the table shows 2 (or more) the microcontroller also has hardware I^2C or SPI capabilities.

PIC	Program (words)	RAM (bytes)	EEPROM (bytes)	I/O pins	Analog input	RS232	USB	I²C	SPI	PWM	CAN bus	Speed (mips)
10f200	256	16	0	4[48]	0	0[49]	0	0	0	0	0	1
12f675	1024	64	128	6	4	1	0	1	1	0	0	5
16f628	2048	224	128	16	0	2	0	1	1	1	0	5
16f628A	2048	224	128	16	0	2	0	1	1	1	0	5
16f648A	4096	256	256	16	0	2	0	1	1	1	0	5
16f88	7168	368	256	16	7	2	0	2	2	1	0	5
16f876A	8192	368	256	25	5	2	0	2	2	2	0	5
16f877	8192	368	256	33	8	2	0	2	2	2	0	5
16f877A	8192	368	256	33	8	2	0	2	2	2	0	5
18f242	8192	768	256	36	8	3[50]	0	2	2	2	0	10
18f2450	8192	768	0	23	10	2	1	1	1	1	0	12
18f4450	8192	768	0	34	13	2	1	1	1	1	0	12
18f4455	12288	2048	256	35[51]	13	2	1	2	2	2	0	12
18f4550	16384	2048	256	35	13	2	0	2	2	2	0	12
18f4685	49152	3328	1024	36	11	2	0	2	2	2	1	10

Table 49. Properties of the supported microcontrollers.

Because of the _bert library all microcontrollers expect to be connected to a 20 MHz crystal except the 10f200 and the 12f675.

---

[48] Pin a3 input only, pin a2 only after the command option_reg = 0b1101_1111.
[49] No software serial capabilities due to a lack of memory.
[50] Including RS485
[51] Pin c3 does not exist, pins c4 and c5 input only.

Wisp connection

The next table applies to the Wisp628 and the Wisp648 programmers. The Wisp648 has a jumper for use with microcontrollers where the MCLR pin must be low before power is applied, to get the microcontroller in programming mode. The Wisp628 requires the use of a special dongle for this. Of all microcontrollers in this book this only applies to the 12f675. The 10f200 is not automatically recognized by the programmer. You need to add "target 10f200" to the command line (target 10f200 port com1 wait err go %F). In JALedit you can do that using Tools, Environment Options and then Programmer.

PIC	yellow	blue	green	white	red	black	jumper
10f200	8	5	4	*	+ 5V	0	no
12f675	4	7	6	*	+ 5V	0	yes
16f628	4	13	12	10	+ 5V	0	no
16f628A	4	13	12	10	+ 5V	0	no
16f648A	4	13	12	10	+ 5V	0	no
16f88	4	13	12	9	+ 5V	0	no
16f876A	1	28	27	24	+ 5V	0	no
16f877	1	40	39	36	+ 5V	0	no
16f877A	1	40	39	36	+ 5V	0	no
18f242	1	28	27	26	+ 5V	0	no
18f2450	1	28	27	26	+ 5V	0	no
18f4450	1	28	27	26	+ 5V	0	no
18f4455	1	40	39	38	+ 5V	0	no
18f4550	1	40	39	38	+ 5V	0	no
18f4685	1	40	39	38	+ 5V	0	no

Table 50. Wisp connections for the supported microcontrollers.

* this wire is not required

12.2 Migration

Migration means that you take a program written for one microcontroller and convert it to another microcontroller. Normally speaking this is rather complicated, but the _bert libraries simplify it quite a bit. As long as you use one of the supported types of course. Naturally you can use other microcontrollers as well but then you will not have the added bonus of the _bert libraries. In practice the range is wide enough for most applications.

12.2.1 How does it work

When you are done with the development of your project you may realize that you might have used a smaller microcontroller. Of perhaps you are only halfway through and already you run out of memory or pins, so you want to move to a larger microcontroller. Personally I always develop on a large microcontroller - usually the 16f877A or the 18f4455 - as you may have noticed in this book. For commercial applications my migration normally consists of moving to a smaller - and thus cheaper - microcontroller.

Before you can start migrating fill out the next table, and search for a microcontroller in the previous tables that meets your requirements.

Item	Required quantity
Program (Flash) memory	
Data memory	
Number of pins	
Special properties	

Migration starts with changing the name of the _bert library. Next you rename pins and optionally change registers. With the latter JAL will help you. You simply compile the program and see if any of the registers generate an error message. This is a fast and easy way to spot issues in the source.

12.2.2 Case 1 - from a 16f877A to a 10f200 (purpose: reduce cost)

As an example let's migrate the tutorial program that runs on a large 16f877A. The program itself is tiny and uses only two pins. When the program is compiled JAL presents us with the following analyses:

jal 2.4j (compiled Mar 12 2009)
0 errors, 0 warnings
Code area: 153 of 8192 used
Data area: 18 of 368 used
Software stack available: 96 bytes
Hardware stack depth 2 of 8

We will use these data to fill in the next table. As number of pins we enter 2, because all the other pins in use are required by the microcontroller itself and not by our program

Item	Required quantity
Program (Flash) memory	153
Data memory	18
Number of pins	2
Special properties	none

In the overview of the supported microcontrollers in section 12.1 you can see that the 10f200 appears to meet our criteria. In fact all criteria are met except for data memory where it is two bytes short. The program uses no variables so the data memory must be used by libraries that the 16f877A_bert library loads by default. In the 10f200 there is no library for software RS232 - because it doesn't have enough memory for that - so it seems reasonable to assume that because of this the 10f200_bert library uses less data memory. So we will first try the 10f200. If that fails we will take the next larger microcontroller in the list: the 12f675. This will always do, but the 12f675 is of course much more expensive: 2.30 euro versus 0.80 euro for the 10f200 - if purchased in bulk.

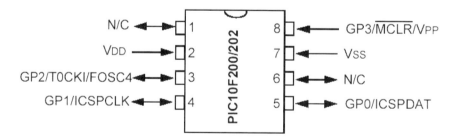

Figure 252. Pin layout of the 10f200 microcontroller.

The first step is to change the library from 16f877A_bert to 10f200_bert. The tutorial program uses two pins, c1 and c2, which the 10f200 doesn't have. So we will rename those pins to a0 and a1. The next table shows both program, the changes are printed bold.

Tutorial program 16f877A	Tutorial program 10f200

-- JAL 2
include 16f877A_bert
 	
-- definitions
pin_c4_direction = Output
pin_d3_direction = Output
 	
forever loop
 	
   -- LEDs in position 1
   pin_d3 = high
   pin_c4 = low
   delay_1s(1)
 	
   -- LEDs reversed
   pin_d3 = low
   pin_c4 = high
   delay_1s(1)
 	
end loop
	-- JAL 2.4j
	include **10f200_bert**
	
	-- definitions
	pin_**a0**_direction = Output
	pin_**a1**_direction = Output
	
	forever loop
	
	   -- LEDs in position 1
	   pin_**a0** = high
	   pin_**a1** = low
	   delay_1s(1)
	
	   -- LEDs reversed
	   pin_**a0** = low
	   pin_**a1** = high
	   delay_1s(1)
	
	end loop

Table 51. Migration of a program.

We compile the program to see if the problem with data memory indeed exists:

    jal 2.4j (compiled Mar 12 2009)
    0 errors, 0 warnings
    Code area: 83 of 256 used
    Data area: 11 of 16 used
    Software stack available: 3 bytes
    Hardware stack depth 2 of 2

No problem at all! In fact we have 5 bytes of data memory available. And because there is no software RS232 connection we also have plenty of program memory left over. That means we can now build the hardware.

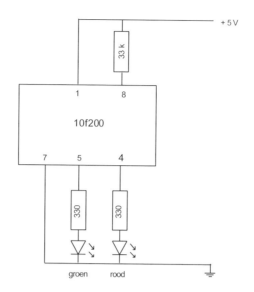

Figure 253. Schematic of the tutorial project.

Figure 254. The tutorial project with a 10f200 on a breadboard.

This tiny microcontroller is not capable of identifying itself to the Wisp programmer. So we need to help it a little bit by telling the Wisp what type it is. You can do that in JALedit using these steps:

1. Go to the menu Tools and then select Environment Options.

2. Select programmer and change this line

    port com1 wait err go %F

    into this line

    target 10f200 port com1 wait err go %F

3. Close the windows and compile and download as usual.
4. Do remember to remove "target 10f200" when you want to program another microcontroller!

12.2.3 Case 2 - from a 16f877A to a 18f4455 (purpose: add USB)

In this second case we again take the tutorial project, but this time we will pretend that you want to expand this program, and more specifically: add USB communications. So you need a microcontroller that has at least the same capabilities as the 16f877A, but with USB. You are not done developing yet so you do not want to go back in capabilities just yet.

The table in section 12.1 shows that the 18f4455 microcontroller has the same amount of Flash, data an EEPROM memory as well as USB.

A footnote at the table shows that the 18f4455 doesn't have pin c3, and that pin c4 and c5 can be inputs only. The tutorial project uses c4 as output so we will change that to c1. Of course we also change the library to 18f4455_bert. In the next table the original and migrated program are listed next to each other. Changes are shown in bold print.

Tutorial program 16f877A	Tutorial program 18f4455

-- JAL 2
include 16f877A_bert
 	
-- definitions
pin_c4_direction = Output
pin_d3_direction = Output
 	
forever loop
 	
   -- LEDs in position 1
   pin_d3 = high
   pin_c4 = low
   delay_1s(1)
 	
   -- LEDs reversed
   pin_d3 = low
   pin_c4 = high
   delay_1s(1)
 	
end loop
	-- JAL 2
	include **18f4455_bert**
	
	-- definities
	pin_**c1**_direction = Output
	pin_d3_direction = Output
	
	forever loop
	
	   -- LEDs in position 1
	   pin_d3 = high
	   pin_**c1** = low
	   delay_1s(1)
	
	   -- LEDs reversed
	   pin_d3 = low
	   pin_**c1** = high
	   delay_1s(1)
	
	end loop

Table 52. Migration of a program.

After this migration the program can be expanded further, and USB functionality can be added as demonstrated in sections 9.5 to 9.7.

13 Appendix

This is the reference guide part of the book. It contains a few unique overviews not found anywhere else. For this reason the book will find its place next to your PC even after the projects have long been built.

13.1 JAL

13.1.1 General

JAL (Just Another Language) is a free-format language for programming PIC microcontrollers. The commands can be spread out over the lines as you please. Tabs, spaces, and newlines are all considered whitespace. There is no delimiter between different commands. In theory you could put all commands on one long line. In practice, however, appropriate whitespace results in easier-to-read programs because any convention you choose is possible.

JAL is the only advanced free language, and has a large and active international user base. It is configurable and extensible by use of libraries and can even be combined with assembler.

A typical JAL program will start with a call to a library that contains relevant details for the microcontroller for which the program is written. If for example you use the 16f877A microcontroller you will use the 16f877A_bert library. Note that the _bert libraries are available only for the microcontrollers that are discussed in this book.

```
include 16f877A_bert
```

Then the variables and constants will be declared:

```
var byte a
```

Next are the commands:

```
forever loop
    a = a + 1
end loop
```

It is a good practice to add comments to your program to indicate what it does and for which JAL version it is written. So a simple program might look like this:

```
--JAL 2.4j
include 16f877A_bert

var byte a

--demo program
forever loop
    a = a + 1
end loop
```

13.1.2 Syntax

Variables

Here the power of JAL is immediately clear. Both unsigned (positive) as well as signed (positive and negative) variables can be used, containing up to 32 bits[52].

name	description	range
bit	1 bit unsigned boolean	0 or 1
byte	8 bit unsigned	0 to[53] 255
sbyte	8 bit signed	-128 to 127
word	16 bit unsigned	0 to 65,535
sword	16 bit signed	-32,768 to 32,767
dword	32 bit unsigned	0 to 4,294,967,296
sdword	32 bit signed	-2,147,483,648 to 2,147,483,647

Table 53. Different types of variables and their ranges.

You can even define variables with any bit length you want, such as:

```
var bit*2 demo
```

Variable *demo* is now 2 bits long (and can thus contain the values 0, 1, 2 or 3). When a value is assigned to *demo* it doesn't necessarily have to be two bits, but only the lowest two bits will be put into the variable (the others do not fit).[54]

---

[52] Variables with a size of more than 16 bits can only be used in microcontrollers that have so called shared memory. Of the microcontrollers in this book the 10f200 and the 12f675 do not have such memory locations, so they are limited to 16 bits or less.
[53] "To" in this context means "up to and including".

So the statement:

>   demo = 99

will result in a value of 3 for *demo*, because the number 99 in binary is 0b_0110_0011, and the lowest two bits are set to 1, which equals 3 in decimal.

Besides decimal values you can use other number bases as well. In that case you need to add a prefix to the number. Possible bases are:

>   23   decimal
>   0x1F   hexadecimal
>   0q07   octal
>   0b01   binary

And of course you can use letters:

>   "Hello"   string

For readability purposes underscores can be added to numbers in any position you want. To the compiler the number 10_0 is identical to 100. Binary value assignments almost always use underscores to make them easier to read:

>   a = 0b_1001_1110

Declaring variables must be done before they can be used.

---

[54] The compiler will notice and give you a warning. If you did it intentionally, you can ignore the warning. If you use the settings in the download package you will not see the warnings, because they are switched off by default.

Here are a few possibilities to do this:

command	description
var byte a	*a* is declared as a byte
var byte a = 0x1F	*a* is declared as a byte and assigned a value at the same time
var byte a at 0x06.	*a* is declared as a byte at memory location 0x06 [55]
var byte volatile a	*a* is declared as a byte and will not be optimized away by the compiler[56]
var byte a is b	*a* is a synonym or alias for *b* (*b* must be declared first)

Table 54. How to declare a variable.

The final declaration can be used to give pins names that are easier to remember. Lets suppose that a red LED is connected to pin c1. If you use this command you can refer to pin c1 using the *redled* alias:

    var bit redled is pin_c1

For example, *redled = 1* would make pin c1 high and thus turn on the LED. This will make the program easier to read. But it is also easier if you want to migrate your program to another PIC. If this microcontroller doesn't have a pin c1 all you have to do is change the declaration to a pin that the microcontroller does have, such as:

    var bit redled is pin_a1

---

[55] It is important to understand what you are doing when you force variables to a certain memory location. It can also be used as a convenient way to split variables. This example declares an array in exactly the same memory location as the word variable *demo*:
    var word demo
    var byte dem[2] at demo
Demo is a word (two bytes). Since the array *dem* is also two bytes long, and in the same memory location, this means that the first item in the array, *dem[0]*, is the low byte and the second item, *dem[1]*, is the high byte.

[56] The compiler optimizes the program during compiling, and will remove all unused parts. References to a pin will also be removed since the compiler doesn't know that pins can change "from the outside". So, to the compiler the command seems useless. The *volatile* keyword is used to prevent this. The compiler will leave a *volatile* variable alone. The average user will almost never need the *volatile* keyword.

The rest of the program can remain unchanged.

Mixing different variable types in a calculation is possible, but care should be taken that intermediate results will be stored in the variable type that the calculation started with. If you multiply a byte with a byte and put the result in a word, like this:

> var byte num1, num2
> var word num3
>
> num3 = num1 * num2

the result will never be larger than 255. This is because *num1* and *num2* are multiplied using bytes and the result is transferred into *num3*.

You can force the compiler to use a word for intermediate calculations by changing the multiplication to *num3 = word(num1) * word(num2)*.

Constants

When you know in advance that a variable will be assigned a value once and will never change it is not a variable but a constant. This can be declared using *const*, like this:

> const byte demo = 5

The advantage of using constants is that a variable uses RAM memory and a constant doesn't (the compiler uses the value rather than the constant name). So it is a good idea to use constants whenever you can. In small programs such as in this book it is not really an advantage, because we are never even close to running out of memory.

Forever loop

This command ensures that a particular part of the program is executed forever. Many microcontroller programs use this command since it is a convenient way to make sure the program never stops.

> forever loop
>   [commands]
> end loop

You can use the command Exit Loop to leave this loop.

While loop

The *while* loop executes a series of commands as long as a certain condition is met (true).

This loop is executed as long as the condition *a* is smaller than *b* is true. If the condition is not met (false) the program will exit the loop.

> while a < b loop
>   [commands]
> end loop

You can use the command Exit Loop to leave this loop even if the condition is still true.

For loop

The *for* loop executes a series of commands a fixed number of times.

This loop is executed ten times:

> for 10 loop
>   [commands]
> end loop

Normally the compiler will count "internally", but you can force it to use a certain variable to count the loops, with the *using* command. This is very convenient if you want to know inside your loop how many times it has already been executed. For example, when you want to determine the position in an array.

> for 10 using counter loop
>   [commands]
>   value = demo[counter]
> end loop

Because microcontrollers start counting at zero the variable counter will get values from 0 to 9. You can use the command Exit Loop to leave this loop even if the loop hasn't been executed enough times.

Repeat until

With this command you can repeat a series of commands until a certain conditions is met (true).

>   repeat
>   [commands]
>   until condition

The next example will repeat the addition until variable a has value 8:

>   repeat
>   a = a + 1
>   until a == 8

Note the double = sign for this is a condition and not a calculation. You can use the command Exit Loop to leave this loop even if the condition is still false.

Procedure

A procedure is a part of program that is needed more than once. Rather than typing it several times it is put aside and given a name. This particular part of the program is only executed when it is "called" by that name. Using procedures usually makes a program easier to read and maintain.

This is an example procedure *demo*

>   procedure demo is
>   [commands]
>   end procedure

You can call this procedure simply by using the name as a command. For the procedure shown above the call would be

>   demo

In procedures you can use variables just as in any other part of the program. Any variable declared outside the procedure is also valid inside a procedure. If you define a variable inside a procedure you can only use it inside that particular procedure. This is called a "local variable".

```
procedure demo is
    var byte a
    [ commands ]
end procedure
```

In this example variable a is declared inside the procedure so it is not known to the rest of the program.

If you want to give a value to a variable from outside the procedure you need to "pass" it to the procedure. The procedure would look like this:

```
procedure demo (byte in a) is
    [ commands that use variable a ]
end procedure
```

Note that this declares variable a. You must not declare it in the main program or in the procedure. Inside the procedure you can now use variable a, which gets a value when the procedure is called. If for example you want to call this procedure and give a value 6 you use this command:

```
demo (6)
```

In this same way you can pass variables out of the procedure, but then you need to declare them as "*byte out*" instead of "*byte in*" in the procedure name. This is an example of a real procedure:

```
procedure Serial_HW_Write ( byte in Data ) is
    -- wait until TXREG is empty
    while ! TXIF loop end loop
    -- then put new byte in TXREG
    TXREG = Data
end procedure
```

If this procedure is for example called using Serial_HW_Write(6) then Data gets value 6, and is used in the line TXREG = DATA.

This is a good way to make new commands that you can add to JAL. If you have made new procedures, for example to control a certain component or function, you can put them in a separate file. This file is then called a library and you can use it by "including" it in your program using the *include* command.

The advantage is that your program becomes much easier to read, and the next time you need that particular component or function you can simply load your library, and off you go.

Interrupt

An interrupt routine is a procedure that is called under certain special conditions. For example when a pin changes value of a timer overruns. The normal program is interrupted (hence the name) and this special procedure is executed. Once the procedure is terminated the normal program will continue where it left off.

```
procedure demo is
    pragma interrupt

    [ command ]

end procedure
```

The first line of an interrupt procedure is the pragma interrupt command. You can have as many interrupt procedures as you want, the compiler will combine them for you. In general however you should try to avoid using interrupts unless it is absolutely necessary.

Interrupt routines may not be called by the main program or other procedures or functions.

Function

A function is basically the same as a procedure. The main difference is that a function always returns a value. The returned value needs to be declared using the *return* statement.

In this function variable *a* is incremented by one:

```
function demo (byte in a) return byte is
    var byte b
    b = a + 1
    return b
end function
```

In the declaration of the function it is indicated that an input is expected (*byte in*) and that the answer the function will return is a byte (*return byte*). Inside the function the "*return b*" statement indicates that *b* will be the value that is returned.

This is an example of a function call:

x = demo(4)

where x will get the value 5 (4 + 1).

Functions are often used to return a status rather than a number, such as *true* or *false*, or a number and a status indication. This is a good example of such a function:

```
function Serial_HW_read ( byte out Data ) return bit is
   -- test if byte available, and if so get byte
   if RCIF then
      Data = RCREG
   else
      return false
   end if

   return true
end function
```

This function first checks to see if RCIF is true, which means that something has arrived in the buffer of the serial port. If true the content of the buffer is copied into variable Data. The function itself gets status true.

If the buffer is empty the function itself gets status false. In that case you know that no information has arrived, and that Data contains no relevant information. If you want your program to wait until information has arrived you can use this line:

while !serial_hw_read(data) loop end loop

As long as the function is false (not true) meaning the buffer was empty, the loop is repeated. As soon as the function is true the loop is exited and Data contains the newly received information.

In-line assembly

You can use assembler inside your JAL programs. I don't think this will ever be really necessary, but perhaps you found a nice snippet on the Internet or want to use an example from the datasheet.

You can use individual assembler statements using the *asm* prefix, like this:

> asm movlw 0xFF

If you need multiple statements it is easier to use an *assembler* block, like this:

> assembler
> > movlw 0xFF
> > bsf 5
> > movwf pr2
> > bcf 5
> end assembler[57]

When your program is compiled the JAL compiler generates a HEX file for the microcontroller as well as an assembler file. You can use this to exchange programs with assembler users or to use Microchip tools.

All assembler commands ("mnemonics") that you find in the datasheet of the microcontroller in your project can be used. Additionally, a few assembler macros that are often used on the Internet can be used too, for example:

command	description
OPTION k	Copy literal k to the *OPTION* register [58]
TRIS {5,6,7}	Copy W to the TRIS {5,6,7} register
MOVFW f	Copy f to W (a synonym for MOVF f, W)[59]
SKPC	A synonym for BTFSS _status, _c
SKPNC	A synonym for BTFSC _status, _c
SKPZ	A synonym for BTFSS _status, _z
SKPNZ	A synonym for BTFSC _status, _z

---

[57] A very complicated way to assign value 0xFF to pr2 in assembler, because the program is in memory bank 0 and pr2 is in memory bank 1. In JAL you don't have to worry about things like that, in fact you don't even need to know what memory banks are. So you can just use pr2 = 0xFF.
[58] The official explanation is *move*. However since the original version is retained it is in fact *copy*.
[59] W is the working register, f is the register file address. *MOVFW myvar* puts the value of *myvar* in W, *MOVLW myvar* puts the address of *myvar* in W.

If you use variables from the JAL program in your assembler block you cannot be sure where they are because JAL may store them in any available memory bank. This also means that at the moment the JAL program is interrupted for your assembler block, you have no clue where the bank pointers are pointing to. That means you don't have access to the registers, even though you do know where they are.

The solution is to ask the compiler to fix the bank pointers for you. You can do this by adding *bank* to every line in which you use JAL variables or registers.

```
procedure register_write( byte in Address, byte in Data ) is
  assembler
     bank  movf address,w  ; move address to working variable w
           movwf FSR       ; move address to file select register
           bcf irp         ; make irp zero so we use bank 0 and 1
                           ; (indirect register bank select), 0 = bank 0/1
     bank  movf data,w     ; move data to w
           movwf indf      ; move data to indirect file register as
                           ; referenced by fsr
  end assembler
end procedure
```

I wrote this fragment for the regedit[60] library and you see that the bank statement is used in all lines containing *address* and *data,* because the location of these JAL variables is unknown. The line referring to the registers FSR, IRP and INDF do not have the bank keyword, because they are accessible from all banks.

Task

Even though a microcontroller can only do one thing at a time you can still do multitasking, just like on a PC. The different tasks can be defined using the *task* command. Every time the program encounters a *suspend* command the current task is interrupted. The scheduler then checks to see which task has been waiting for the longest time and hands control over to that task. This system only works if the tasks are "honest" and use the *suspend* command regularly.

---

[60] This is a library to read and change registers while the program is running. This is not discussed in this book. The latest version of the library (contained in the download package) is written in JAL, so the fragment is no longer used.

Just like in a procedure, tasks can have their own local variables.

```
task name1 (parameters) is
[ commands ]
    suspend
    [ commands ]
end task

task name2 (parameters) is
[ commands ]
    suspend
    [ commands ]
end task

start task1(parameter values)
start task2(parameter values)

forever loop
    suspend
end loop
```

Before a task can be run you need to start it using *start*. The task can be stopped using *end task*.

IMPORTANT: If you want to compile a program containing tasks you need to tell the compiler how many tasks there are. Note that the main program is also counted as a task. So for the previous example the program should contain this command:

Pragma task 3

In older JAL versions the number of tasks must be submitted in the command line, so in this example using -task 3[61].

If then else

This command is used to make a choice between two possibilities. *If* one condition occurs *then* something is done, *else* something else is done.

In this example *b* gets the value of 2 when *a* is equal to 1. In all other cases *b* gets 3.

---

[61] If you use JALedit you can enter this at Compile - Environment Option - Additional Commandline Parameters.

```
if a == 1 then
    b = 2
else
    b = 3
end if
```

This command can be nested, like in this example.

```
if a == 1 then
    b = 2
else if a == 2 then
        b = 3
    else
        b = 4
    end if
end if
```

Note that *else if* are two words. The above program yields the following results:

if	then
a = 1	b = 2
a = 2	b = 3
a = something else	b = 4

Instead of *else if* you can also use the command *elsif*. The syntax is as follows:

```
if a == 1 then
    b = 2
elsif a == 2 then
    b = 3
elsif a == 3 then
    b = 4
else
    b = 5
end if
```

The *elsif* statement is not used very often.

Case

If you need to make multiple decisions based on the value of a single variable the *case* statement is a good alternative to a row of *if then else* statements. In the following example *b* is given a value based on the value of *a*. If *a* is 3 than *b* will be 4. If the value of *a* is not listed then *b* will get the value 6.

```
case a of
    1: b = 20
    2: b = 13
    3: b = 4
    4: b = 59
    otherwise b = 6
end case
```

Contrary to the *if then else* statement the *case* block can only have one statement per choice. So if you want to give *b* the value 13 and *aa* the value 188 when *a* is 2 then this is not possible. You will have to use a *procedure*, or the *block* statement.

Block

The *block* command can be used to group statements together. Variables defined inside a *block* can only be used in this *block*. A block is a program section and will be executed when it is encountered. In cannot be called form another location.

```
block
    [ commands ]
end block
```

This is particularly useful in combination with the case command, or if you want to keep variables local to just a small portion of your program. This could happen if you want to use parts of different programs combined into one, that all use the same variable names.

```
case a of
    1: b = 20
    2: block
          b = 13
          c = 188
       end block
    3: b = 4
    otherwise b = 6
end case
```

Array

Normally speaking a variable has only one value. With an array a variable can be given a whole range of variables. In the following example the *demo* array is assigned a row of five values.

    var byte demo[5] = {11,12,13,14,15}

To get a value out of the array the number between square brackets indicates the position that you want. Remember that computers start counting at 0, so the first position in the array is 0. In our example *demo[0]* contains the value 11.

This command selects the fourth number in the array (the value 14):

    a = demo[3]

Adding a value to an array (or modifying one) is done in a similar way. In this command the fourth position in the array is assigned the value in *a*:

    demo[3] = a

Your program can use the *count* statement to determine the length of an array. For example:

    a = count(demo)

That means that count is a reserved word and cannot be used as a variable name. Be careful with your array size, since an array has to fit within a single RAM memory bank. A bit array is not possible.

Long table (also known as: Lookup table or LUT)

If you need more space that you can fit into an array you can use the long table. This is an array with just constants. That means you define it once, and afterward you cannot make changes to its content. This is because even though it is an array, it is not contained in RAM, but in program (flash) memory.

The long table can be very long indeed; in the 16F877A a whopping 8000 bytes! Of course the length can never be more than the available program memory. So in a small PIC, or with a large program, long tables must be shorter. You can check this by compiling the program with a long table with a length of 1. Based on the resulting program size the available length for the long table can then be calculated. See section 6.3 for more information on how to determine the size of a program.

If the long table doesn't fit the compiler will generate an error, and your program will not be downloaded into the 16F877A.

The syntax of the long table is:

>   const byte long_table[2049] = { 1, 2, 3, 4, 5, ...}
>   var word x
>   var byte y
>
>   y = long_table[x]

where the above values are just for demonstration purposes.

Operators

JAL has a wide variety of operators. The most important ones are shown below, with a simple example if appropriate.

Operator	Explanation
!!	Logical. Indicates whether a variable is zero or not. For example !!5 = 1, and !!0 = 0
!	Not, or Complement. Flips 0 to 1 and 1 to 0 at the bit level. So !5 = 250 because !0b_0000_0101 = 0b_1111_1010
*	Multiply.
/	Divide without remainder. So 9/2 = 4
%	The remainder of the division, also known as MOD. So 9%2 = 1
+	Add.
-	Subtract.
<<	Left shift. Move all bits one position to the left. Note that the newly created bit is set to 0. When a signed variable is shifted the sign is retained.

Operator	Explanation
>>	Right shift. Same as left shift, but in the other direction.
<	Less than.
<=	Less than or equal to.
==	Equal to. Note that these are two equal signs == in a row. Accidentally using only one = is a very common mistake (feel free to view this as an understatement). The single = sign is a calculation not a condition.
!=	Not equal to.
>=	Greater than or equal to.
>	Greater than.
&	AND comparison at the bit level. The truth table is: 1 & 1 = 1 1 & 0 = 0 0 & 1 = 0 0 & 0 = 0
|	OR comparison at the bit level. The truth table is: 1 | 1 = 1 1 | 0 = 1 0 | 1 = 1 0 | 0 = 0
^	XOR (eXclusive OR) comparison at the bit level. The truth table is: 1 ^ 1 = 0 1 ^ 0 = 1 0 ^ 1 = 1 0 ^ 0 = 0

Table 55. Operators.

Pragma

Pragma is a command for the compiler. It is a very powerful but complex command. A detailed explanation would be out of scope for this book, particularly since you will never (or rarely) use most of the commands.

This table lists the most important *pragma*'s:

command	description
pragma eedata	Stores data in the EEPROM of the microcontroller. For example *pragma eedata* "O","K" will store the letters O and K.
pragma interrupt	This can only be used inside a procedure. The result is that the procedure is added to the interrupt chain, a series of procedures that is executed when an interrupt occurs. There is no limit to the number of procedures in the chain, but the execution order is not defined. Interrupt procedures may not be called from the program itself of from other proecdures or functions.
Pragma clear yes	Give all variables value zero at the beginning of the program. This does not apply to volatile variables or variables that were given an explicit value by the programmer.
Pragma task	Indicates the number of tasks in the program. If you use tasks then remember that the main program counts as a task too. You can either use this command or enter the number of tasks on the commandline. This last option is mainly for older versions of JAL that don't support this pragma.

Table 56. A few of the most relevant pragma's.

Comments

Lines containing comments are preceded by two dashes or a semicolon. You need to do this for each line, as there is no block comment statement.

    ; this is a comment
    -- and this too

Comment lines are used to clarify what a program is for, or why things are done the way they are done. This is very handy for future reference, or when you want to share programs over the Internet.

It is good practice to indicate the JAL version on the very first line of your program. That eliminates a lot of questions!

Good comments are not about what a certain statement does (unless you are writing a book), because the reader may be expected to know this. They are about <u>why</u> you use the statement. If you make a library you should use comment lines to explain in detail what the library is for and how it should be used. Libraries without these comments are completely useless.

13.2 Library _bert

Libraries are used to keep the specific microcontroller settings, registers and variables easily accessible, and to add extra commands to JAL. Because everyone can write and publish libraries, not all of them work will together.

With the free download that comes with this book you will find a series of libraries that have been combined into one big library for each microcontroller, such as the 18f4455_bert library. These libraries add a wide range of extra commands to JAL. Besides, all compatibility problems have been fixed. The credit for the individual libraries within this pack goes to the individual writers.

This combination library (the "standard library") adds commands to JAL that are used in many different parts of the book, such as analog to digital conversion, serial communication, reading and writing memory, and much much more. The most important ones are explained in this section. Of course these are only applicable if the microcontroller is actually equipped with the hardware to support these functions. In section 12 you will find an overview off all microcontrollers for which a _bert library exists, and which functionality is applicable.

Serial communication

All microcontrollers in can make use of serial communication using software emulation - with the exception of the 10f200 which is too small. This is the technique that is used in many places in this book when the Wisp648 is used to connect the microcontroller to a PC using the passthrough functionality of the programmer. Contrary to other types of communication software serial does not have any buffers. So when a signal is coming in while the microcontroller isn't actually waiting for it, it will be lost. You need to be aware of this when setting up communications.

command	description
Serial_SW_Baudrate	Set the baudrate (communications speed) for the RS 232 connection. Software communication speed in the standard library is set to 1200 by default, but other speeds such as 19k2 or 115k baud are also possible.
Serial_SW_Invert	Invert the signal.
Serial_SW_Write_Init	Prepare to send data from the microcontroller over the RS232 connection.
Serial_SW_Read_Init	Prepare to receive data over the RS232 connection.
Serial_SW_Read(data)	Receive data and put it in the variable *data*.
Serial_SW_Write (data)	Send the contents of the variable *data*.
Serial_SW_Locate(horizontal, vertical)	Move the cursor (on a VT52 terminal or emulation) to the coordinates horizontal, vertical.
Serial_SW_Clear	Clear the screen from the current cursor position (on a VT52 terminal or emulation).
Serial_SW_Home	Move the cursor to the home position - the upper left corner (on a VT52 terminal or emulation).
Serial_SW_Byte(data)	Send a byte named data as three digits - or less - instead of a single number .
Serial_SW_Printf(array)	Send a complete array - which needs to be defined first - with a single command. For example const byte mystr[] = "Bert van Dam" followed by serial_sw_printf(mystr).

Many of the microcontrollers discussed in this book have a serial communications unit built in. For this hardware serial communication the following commands apply:

command	description
USART_HW_Serial	Set the right protocol: true = RS232, false = SPI (default is true). Since a different library is used for SPI it is best not to change the default setting.
Serial_HW_Baudrate	Set the baudrate (communication speed) for the RS232 connection. Hardware communication speed in the standard library is set to 19k2 by default, but other speeds such as 1200 or 115k baud are also possible.
Serial_HW_Read(data)	Receive serial data and store it in the variable *data*. You can use this command also to check if any data has arrived in the buffer, see for example section 10.1
Serial_HW_Write(data)	Send the contents of the variable *data*.
Serial_HW_Data = data	The same as above, but called as a function.
Serial_HW_Locate(horizontal, vertical)	Move the cursor (on a VT52 terminal or emulation) to the coordinates horizontal, vertical.
Serial_HW_Clear	Clear the screen from the current cursor position (on a VT52 terminal or emulation).
Serial_HW_Home	Move the cursor to the home position - the upper left corner (on a VT52 terminal or emulation)
Serial_HW_Byte(data)	Send a byte named data as three digits - or less- instead of a single number.
Serial_HW_Printf(array)	Send a complete array - which needs to be defined first - with a single command. For example const byte mystr[] = "Bert van Dam" followed by serial_hw_printf(mystr).

Pulse Width Modulation (PWM)

The PWM module of the microcontroller is used, so these commands can only be used for microcontrollers that have a PWM module.

command	description
const pwm_frequency = number	This sets the PWM frequency. You can only chose the frequencies listed in the datasheet and as you have seen this has consequences for the available resolution.
const pwm1_dutycycle = var const pwm2_dutycycle = var	If you use frequency modulation (FM) instead of PWM this is where you set the desired duty cycle.
PWM_init_frequency (boolean,boolean)	Initialize the PWM modules.
PWM_Set_DutyCycle (var,var)	Set the duty cycle. When using PWM this is the variable that controls the power output (if an electric motor is used this controls the speed).

When a microcontroller is used that only has one PWM module, you can use the same commands. This makes migration from one microcontroller to another simple. Values for the second PWM module will be ignored. To allow migration to dual PWM microcontrollers it is best to use the value 0 for the modules that don't exist. For example:

command	description
PWM_init_frequency (boolean,0)	Initialize PWM module 1.
PWM_Set_DutyCycle (var,0)	Set the duty cycle on module 1. When using PWM this is the variable that controls the power to the user (if the user is an electric motor it controls the speed)

A/D conversion

This library can only be used for microcontrollers that are equipped with A/D capabilities.

command	description
ADC_init	Initialize the A/D converter.
ADC_hardware_Nchan	Set the number of A/D channels you want to use. See the table for the 16f877A channels, for other microcontrollers consult the datasheet.
ADC_on	Switch A/D on (for all channels).
ADC_off	Switch A/D off (all pins become digital).
ADC_read (chan)	Read the analog value on a channel into a word (channel is the AN number in the table).
ADC_read_bytes(chan, Hbyte, Lbyte)	Read the analog value on a channel into two separate bytes.
ADC_read_low_res (chan)	Read the analog value on a channel into one byte, with low resolution (max value: 255).

Using the ADC_hardware_Nchan constant may be confusing due to the lack of logic. Consult the following table to see which pin is used for a specific channel for the 16f877A. Check the datasheet for other microcontrollers.

pin	A0	A1	A2	A3	A5	E0	E1	E2
A/D channel	AN0	AN1	AN2	AN3	AN4	AN5	AN6	AN7
0								
1	a							
3	a	a		a				
5	a	a	a	a	a			
6	a	a	a	a	a	a		
8	a	a	a	a	a	a	a	a

Please note that pin a4 has no A/D converter and that a6 and a7 do not exist (port A is only 6 bits wide). Channel AN0 is connected to pin a0. But, confusingly enough, AN4 is connected to pin a5, and AN7 is connected to pin e2. So take care when developing programs; mistakes are easily made.

Also note the strange configuration when three analog channels are selected. Against all expectations pin a2 is switched off and pin a3 is switched on.

Program memory

This library can only be used with the larger microcontrollers.

command	description
Program_EEprom_Read(Address, Data)	Read the program memory *address* and store the result in *data*. This command is safe; you could even write a program that reads itself.
Data=Program_EEprom(Address)	A different way to achieve the same result.
Program_EEprom_Write(Address, Data)	Writes the value in *data* to program memory location *address*. You can accidentally overwrite your program with this command.[62]

EEPROM memory

This library can only be used with microcontrollers that have EEPROM memory.

command	description
Data_EEprom_Write(Address,Data)	Write the value in *data* to the *address* memory location in EEPROM.
Data_EEprom_Read(Address,Data)	Read the *address* memory location in EEPROM and put the contents into the *data* variable.
Data=Data_EEprom(Address)	The same as above, but written as a function.
Pragma EEDATA data1,data2, etc	Let the compiler store data in EEPROM. No need to mess with addresses, and you can fit a lot of data on a single line.

---

[62] Of course you can also overwrite your program on purpose, which would be a self-modifying program.

Delay

The following delays are available for your programs. The number indicates how often a particular delay is executed. *Delay_100ms(3)* means a delay of 3 x 100 mS = 300 ms.

    delay_1us
    delay_1usM (byte in N)
    delay_2uS
    delay_3uS
    delay_4uS
    delay_5uS
    delay_6uS
    delay_7uS
    delay_8uS
    delay_9uS
    delay_10uS (byte in N)
    delay_20us (byte in N)
    delay_50us (byte in N)
    delay_100us (byte in N)
    delay_200us (byte in N)
    delay_500us (byte in N)
    delay_1ms (byte in N)
    delay_2ms (byte in N)
    delay_5ms (byte in N)
    delay_10ms (byte in N)
    delay_20ms (byte in N)
    delay_50ms (byte in N)
    delay_100ms (word in N)
    delay_200ms (byte in N)
    delay_500ms (byte in N)
    delay_1s (byte in N)
    delay_2s (byte in N)
    delay_5s (byte in N)

Random

This library can be used with all microcontrollers to generate pseudo random numbers.

command	description
number = random_WORD	Generate a random number in the range 0 to 65,535.
number = random_BYTE	Generate a random number in the range 0 to 255.
number = dice	Generate a random number in the range 1 to 6 (specifically meant for dice).

Registers and variables

Microcontrollers contain several registers. Each of these registers is defined in the _bert library. This means that any register you encounter in the datasheet can be directly addressed in your program.[63] The intcon register of the 16f877A, for example, can be set as follows:

intcon = 0b_1010_0000

In many cases individual bits in a register have their own name. In the intcon register they are gie, peie, t0ie, inte, rbie, t0if, intf and rbif (see figure 254). These individual variables are also defined and usable.

So instead of the command:

intcon = 0b_1010_0000

You could write:

gie = 1
t0ie = 1

The difference is that the first option explicitly sets all unused bits to 0, while the second option leaves the unused bits in whatever status they were before. The individual

---

[63] If you use JAL you'll think this is very normal. Just for fun take a look at the 16F877A_inc_all.jal file. This is a list of memory locations and banks of all registers and variables. For the INTCON register this is 0x0B,0x8B,0x10B,0x18B in bank 0. For the GIE variable it is intcon : 7.

variables, therefore, are especially convenient when the entire register is set and you want to change just one bit.

INTCON REGISTER (ADDRESS 0Bh, 8Bh, 10Bh, 18Bh)

R/W-0	R/W-0	R/W-0	R/W-0	R/W-0	R/W-0	R/W-0	R/W-x
GIE	PEIE	T0IE	INTE	RBIE	T0IF	INTF	RBIF
bit 7							bit 0

bit 7   **GIE**: Global Interrupt Enable bit
       1 = Enables all unmasked interrupts
       0 = Disables all interrupts

bit 6   **PEIE**: Peripheral Interrupt Enable bit
       1 = Enables all unmasked peripheral interrupts
       0 = Disables all peripheral interrupts

bit 5   **T0IE**: TMR0 Overflow Interrupt Enable bit
       1 = Enables the TMR0 interrupt
       0 = Disables the TMR0 interrupt

bit 4   **INTE**: RB0/INT External Interrupt Enable bit
       1 = Enables the RB0/INT external interrupt
       0 = Disables the RB0/INT external interrupt

bit 3   **RBIE**: RB Port Change Interrupt Enable bit
       1 = Enables the RB port change interrupt
       0 = Disables the RB port change interrupt

bit 2   **T0IF**: TMR0 Overflow Interrupt Flag bit
       1 = TMR0 register has overflowed (must be cleared in software)
       0 = TMR0 register did not overflow

bit 1   **INTF**: RB0/INT External Interrupt Flag bit
       1 = The RB0/INT external interrupt occurred (must be cleared in software)
       0 = The RB0/INT external interrupt did not occur

bit 0   **RBIF**: RB Port Change Interrupt Flag bit
       1 = At least one of the RB7:RB4 pins changed state; a mismatch condition will continue to set the bit. Reading PORTB will end the mismatch condition and allow the bit to be cleared (must be cleared in software).
       0 = None of the RB7:RB4 pins have changed state

Figure 255 variables in the intcon register

13.3 Other libraries

In this section libraries are discussed that are part of the download package, and are used in this book. You can use these libraries in your own programs with the *include* command. You'll also need the *_bert* standard library. If for example you want to use the usb_rs232 library in the 18f4455 your program would start with the following lines:

include 18f4455_bert
include usb_rs232

Not all libraries are discussed in this book. Some of them have been discussed in detail in a previous book that contains 50 different interesting and exiting projects[64].

USB serial (library usb_rs232)

Some of the microcontrollers in this book are equipped with a USB module. You can choose between three different libraries. For USB serial communication you use the library usb_rs232. This library gives you the following commands:

command	description
Serial_USB_Poll	Take care of USB business and make sure the connection is still open.
Enable_Modem	You can use this variable to check if the USB-serial link is still up.
Serial_USB_Read(Data)	Receive serial data and store it in the variable *data*. You can use this command also to check if any data has arrived in the buffer, see for example section 9.5.
Serial_USB_Write(Data)	Send the content of variable data.
Serial_USB_Locate(horizontal, vertical)	Move the cursor (on a VT52 terminal or emulation) to the coordinates horizontal, vertical.
Serial_USB_Clear	Clear the screen from the current cursor position (on a VT52 terminal or emulation).
Serial_USB_Home	Move the cursor to the home position - the upper left corner (on a VT52 terminal or emulation).
Serial_USB_Byte(data)	Send a byte named data as three digits - or less- instead of a single number.

---

[64] PIC Microcontrollers - 50 projects for beginners and experts

command	description
Serial_USB_Printf(array)	Send a complete array - which needs to be defined first - with a single command. For example const byte mystr[] = "Bert van Dam" followed by serial_usb_printf(mystr).
serial_usb_disable	Disable the USB connection. Do remember to - manually - disable the connection on the PC side as well.
serial_usb_enable	Enable the USB connection if you have disabled it with the previous command. Make sure to wait at least 3 seconds between these two commands. Do remember to - manually - enable the connection on the PC side as well.

USB HID keyboard (library usb_hid_keyboard)

This library is meant for USB communication where your project pretends to be a keyboard.

command	description
usb_is_configured()	Check to see if the USB keyboard link is still up.
usb_tasks	Take care of USB business and make sure the connection is still open.
hid_keyboard_byte(data)	Send a byte named data as three digits - or less- instead of a single number .
hid_keyboard_key(data)	Send data to the PC as byte in a keyboard report (used to send a key to the PC).

USB HID mouse (library usb_hid_mouse)

This library is meant for USB communication where your project pretends to be a mouse.

command	description
HID_Mouse_Write	Send the mouse report to the PC
usb_delay_1ms(n)	Wait n (1 to 255) times 1 ms, while the USB connection is kept active.
usb_delay_100ms(n)	Wait n (1 to 255) times 100 ms, while the USB connection is kept active.
usb_tasks	Take care of USB business and make sure the connection is still open.
usb_is_configured	Check to see if the USB mouse link is still up.
usb_initialized	A flag to remember if usb_tasks needs to run the SUB setup or not.

EEPROM (I^2C) (library i2c_sw)

This library is not used in this book.

command	description
i2c_sw_write(chipaddress,addressH, addressL, value)	Write *value* to *addressH*, *addressL* of the EEPROM having an address of *chipaddress*.
i2c_sw_read(chipaddress,addressH, addressL, value)	Read the contents *(value)* of *addressH*, *addressL* of the EEPROM having an address of *chipaddress*.
i2c_sw_ackpoll(chipaddress)	Wait for an acknowledgement from the EEPROM having an address of *chipaddress*.

This is a software library which means it can also be used with microcontrollers that don't have an I²C module. If you want to use different pins you can modify this section in the i2cp.jal file:

```
var volatile bit i2c_clock_in   is pin_c3
var volatile bit i2c_clock_out  is pin_c3_direction
var volatile bit i2c_data_in    is pin_c4
var volatile bit i2c_data_out   is pin_c4_direction
```

This library is meant for I²C EEPROM chips.

LCD display (library lcd_44780)

Commands in the lcd_44780 library (not used in this book):

command	description
LCD_init	Initialize the LCD display for use (this command is automatically called when you load the library).
LCD_clear_line (line)	Clear line (note: the first line is number 0).
LCD_char_pos (character, position)	Print a character at the position indicated (note: the first position is 0).
LCD_char_line_pos (character, line, position)	Print a character on the position and line indicated.
LCD_num_pos (byte, position)	Print a number (0 to 255) at the position indicated.
LCD_num_line_pos (byte, line, position)	Print a number (0 to 255) at the position and line indicated.
LCD_num_pos_1dec (byte, position)	Print a number (0 to 255) at the position indicated with one decimal digit (so 255 will be printed as 25.5).
LCD_high_low_line_pos (hbyte, lbyte, line, position)	Print a number (0 to 65535) at the position and line indicated.
LCD_progress (byte, line)	Display a bar graph with length of *byte* (maximum 16).

command	description
LCD_shift_right	Move the entire display (both lines) to the right.
LCD_shift_left	Move the entire display (both lines) to the left.
LCD_cursor_pos = position	Place the cursor at the position indicated.
LCD_cursor = off	Switch the cursor off (or "on").
LCD_blink = on	Switch cursor blinking on (or "off").
LCD_display = off	Switch the entire display off (or "on").
LCD_custom(memory address)	Put a custom character in the LCD memory (use only addresses 0 to 7).
CharData[]	Array to define the custom character. Send to the LCD memory one character at a time.
LCD_clock_line_pos (byte, line, position))	Print a number (00 to 99) at the indicated position, uses a leading zero if less than 10.

A typical display layout would be:

line	position															
0	0	1	2	3	4	5	6	7	8	9	10	11	12	13	14	15
1	0	1	2	3	4	5	6	7	8	9	10	11	12	13	14	15

Note that both line and column numbers start at 0.

The *LCD_num_pos(byte,18)* command writes outside the visible area. This is not a problem, but you won't see any of it. If you make the position larger you will eventually end up on the next line.

View and edit registers (library regedit)

This library is meant for use with the REGedit (register editor) program for the PC. This library is not used in this book, nor is the REGedit program part of the download packet.

command	description
procedure register_write(Address, Data)	Write the *data* byte to the register *address*.
procedure register_read(Address, Data)	Read the value of the register at *address* and put it into the *data* variable.
procedure register_debug	Communication procedure for REGedit, which uses a serial connection to observe and modify registers on the fly.

1-wire (library 1_wire)

This is a software library, it can be used on any microcontroller, even those that don't have 1-wire functionality built in.

command	description
d1w_write_bit	Write an individual bit.
d1w_read_bit	Read an individual bit.
d1w_write_byte	Write a byte.
d1w_read_byte	Read a byte.
d1w_reset	Give a reset.
d1w_present	See if anyone is present on the bus.

1-wire DS1882 (library ds1822_1_wire)

You can use this library only if the 1-wire library is loaded first. For example like this:

```
include 18f4455_bert
include 1_wire
include ds1822_1_wire
```

command	description
read_ID	Request the ID number - or address - of a component (slave). There must be only one slave on the bus otherwise they will all reply at the same time.
MatchRom	Attention slaves: the master is about to send an address.
Send_ID	Send the address.
DS1822_start_temperature_conversion	Order the DS1822 to determine the temperature.
DS1822_read_temperature_raw	Read the temperature without checking the CRC (control number). Not recommended.
d1w_read_byte_with_CRC	Read a group of bytes and check the CRC (control number). In order to get the temperature you must read 9 bytes.
Load_My_ID	You can store addresses in the library and retrieve them as needed.

SPI hardware (library spi_hardware)

This library uses the built in SPI capabilities of the microcontroller. It can only be used with microcontrollers that actually have such a unit.

command	description
data0 = spi_transceive(data1)	A function that sends data1 using the SPI protocol and receives data0 at the same time. Can be used in both slave and master.

SPI software (library spi_software)

This is a software SPI library that enables the use of SPI on microcontrollers that have no SPI unit. The master clock frequency is fixed at about 620 kHz.

command	description
data0 = spi_transceive(data1)	A function that sends data1 using the SPI protocol and receives data0 at the same time. Can be used in both slave and master.

MMC card software (library mmc)

This library contains commands that can be used with MMC cards (not SD). Since MMC uses SPI an SPI library must be loaded first. It is recommended you use the software SPI library because it is fast enough for MMC and doesn't require any settings:

    include 16F877A_bert
    include spi_software
    include mmc

command	description
mmc(byte in Cmd,word in dataH,word in dataL,byte in CRC)	Send a command to the MMC card with dataH and dataL as payload, including an CRC.
mmc_init	Initialize the MMC card.
MMC_write_open(word in dataH, word in dataL)	Open the MMC card for a block write operation (in blocks of 512 bytes), with dataH and dataL as payload.
MMC_write_close	Close the block write operation.
MMC_read_open(word out dataH, word out dataL)	Open the MMC card for a block read operation (in blocks of 512 bytes), with dataH and dataL as payload.
MMC_read_close	Close the block read operation.
MMC_read_streaming_open(word in dataH,word in dataL)	Open the MMC card in streaming mode, with dataH and dataL as payload.
MMC_read_streaming_close	Close the streaming mode.

Possible error messages generated by the functions
(use for example error = MMC_write_open(0,0))

    0 = everything ok
    1 = block write open failed
    2 = block write failed
    3 = block read open failed
    4 = streaming mode open failed

Camera routines (library cmucam2)

This library contains routines that are convenient when communicating with the CMUcam2 (or CMUcam3 in cam2 emulation mode).

command	description
wait_for_colon	Wait for the receipt of a colon (:) on the serial hardware connection.
wait_for_packet	Wait for the receipt of a byte with value 255 on the serial hardware connection. This means a packet of data is present.
wait_for_reply(data)	Wait for the receipt of a data on the serial hardware connection and store it in the variable data.

DS1307 real time clock (library ds1307)

This library contains routines for use with the DS130 real time clock.

command	description
ds1307_entertime(sec,min,hr)	Request the time using a PC screen and prepare the answers for the next command.
ds1307_sendtime(sec,min,hr)	Send the time to the ds1307.
ds1307_readtime	Request the time from the DS1307 and prepare the answers for the next command.

command	description
ds1307_showtime	Display the time on the PC screen using the format hh:mm:ss

I²C library (library i2c)

This is a software library that can (also) be used with microcontrollers that have no I²C capabilities. The pins used for I²C are in the i2cp.jal file in the library directory.

command	description
i2c_put_write_address(x)	The master announces that it wants to write to the slave with address x.
i2c_put_data(x)	Send data x.
i2c_put_stop	Send a stop signal.
i2c_put_read_address(x)	The master announces that it wants to receive data from the slave with address x.
i2c_get_data(x)	Receive data.
i2c_put_ack	Send an ack (acknowledge).
i2c_put_nack	Send a nack (not acknowledge).
i2c_wait_ack	Wait for an ack (acknowledge).

7-segment library (library 7segment)

The 7segment library contains all digits from 0 through to 9.

command	description
segnumber(n)	Show the number n on a 7-segment display. See the library for the connections of the segments to the pins.

VT52 commands

For use with VT52 emulation terminal programs.

Command	Result
Esc A	Cursor one line up
Esc B	Cursor one line down
Esc C	Cursor one position to the right
Esc D	Cursor one position to the left
Esc H	Cursor to the home position (top left)
Esc J	Erase everything from the cursor
Esc K	Erase the rest of the line
Esc Y row+32 col+32	Go to location row, col

13.4 ASCII table

Computers use numbers, not letters or other signs. So in order to use these letters and signs it is agreed upon that they are represented by certain numbers. Which letters and signs are represented by which numbers is recorded in the so-called ASCII[65] table. This agreement was made back when computers used 7-bit numbers, therefore only 127 signs were defined. Numbers above 127 can be used, but every manufacturer can make his own choices. These are often used for language dependent signs such as the Euro sign.

If for example you want to send the letter A to the PC you can use this command:

    serial_sw_write(65)

In the terminal program on the PC this will be shown as an A.

JAL know all ASCII values, so you could also use the command:

    serial_sw_write("A")

Note that this still sends 65 to the PC. In the next table you will find an overview of the ASCII codes.

---

[65] American Standard for Computer Information Interchange.

ascii	sign	ascii	sign	ascii	sign
0	ctl@	43	+	86	V
1	ctlA	44	,	87	W
2	ctlB	45	-	88	X
3	ctlC	46	.	89	Y
4	ctlD	47	/	90	Z
5	ctlE	48	0	91	[
6	ctlF	49	1	92	\
7	ctlG	50	2	93	]
8	ctlH	51	3	94	^
9	ctlI	52	4	95	_
10	ctlJ	53	5	96	`
11	ctlK	54	6	97	a
12	ctlL	55	7	98	b
13	ctlM	56	8	99	c
14	ctlN	57	9	100	d
15	ctlO	58	:	101	e
16	ctlP	59	;	102	f
17	ctlQ	60	<	103	g
18	ctlR	61	=	104	h
19	ctlS	62	>	105	i
20	ctlT	63	?	106	j
21	ctlU	64	@	107	k
22	ctlV	65	A	108	l
23	ctlW	66	B	109	m
24	ctlX	67	C	110	n
25	ctlY	68	D	111	o
26	ctlZ	69	E	112	p
27	ctl[	70	F	113	q
28	ctl\	71	G	114	r
29	ctl]	72	H	115	s
30	ctl^	73	I	116	t
31	ctl_	74	J	117	u
32	Space	75	K	118	v

ascii	sign	ascii	sign	ascii	sign
33	!	76	L	119	w
34	"	77	M	120	x
35	#	78	N	121	y
36	$	79	O	122	z
37	%	80	P	123	{
38	&	81	Q	124	|
39	'	82	R	125	}
40	(	83	S	126	~
41	)	84	T	127	DEL
42	*	85	U		

13.5 Keyboard scancodes

This is an overview of three sets of keyboard scancodes that your PC might use. In this book we use set three, but it is possible that your computer uses another set. In section 9.7 you will find instructions on how you can determine which keyboard scancodes your PC uses.

```
Scan Code              Key      Scan Code                    Key
Set    Set   Set  USB           Set    Set    Set  USB
1      2     3                  1      2      3
01     76    08   29   Esc      37     7C                    * PrtSc
02     16    16   1E   ! 1      37+    7C+    7E   55        * KP
03     1E    1E   1F   @ 2      37/54+ 7C/84  57   46        PrtSc
04     26    26   20   # 3      38            11   19   E2   Alt L
05     25    25   21   $ 4      E0 38  E0 11  39        E6   Alt R
06     2E    2E   22   % 5      39            29   29   2C   Space
07     36    36   23   ^ 6      3A            58   14   39   Caps Lock
08     3D    3D   24   & 7      3B            05   07   3A   F1
09     3E    3E   25   * 8      3C            06   0F   3B   F2
0A     46    46   26   ( 9      3D            04   17   3C   F3
0B     45    45   27   ) 0      3E            0C   1F   3D   F4
0C     4E    4E   2D    -       3F            03   27   3E   F5
0D     55    55   2E   + =      40            0B   2F   3F   F6
0E     66    66   2A   Back Space  41         83   37   40   F7
0F     0D    0D   2B   Tab      42            0A   3F   41   F8
10     15    15   14   Q        43            01   47   42   F9
11     1D    1D   1A   W        44            09   4F   43   F10
12     24    24   08   E        45+           77+  76   53   Num Lock
13     2D    2D   15   R        45/46+        77/7E+ 62 48   Pause/Bk
14     2C    2C   17   T        46            7E             ScrLk/Bk
```

Scan Code				Key	Scan Code				Key
Set	Set	Set	USB		Set	Set	Set	USB	
15	35	35	1C	Y	46+	7E+	5F	47	Scroll Lock
16	3C	3C	18	U	47	6C	6C	5F	7 Home KP
17	43	43	0C	I	E0 47*	E0 6C*	6E	4A	Home CP
18	44	44	12	O	48	75	75	60	8 Up KP
19	4D	4D	13	P	E0 48*	E0 75*	63	52	Up CP
1A	54	54	2F	{ [	49	7D	7D	61	9 PgUp KP
1B	5B	5B	30	}]	E0 49*	E0 7D*	6F	4B	PgUp CP
1C	5A	5A	28	Enter	4A	7B	84	56	- KP
E0 1C	E0 5A	79	58	Enter KP	4B	6B	6B	5C	4 Left KP
1D	14	11	E0	Ctrl L	E0 4B*	E0 6B*	61	50	Left CP
E0 1D	E0 14	58	E4	Ctrl R	4C	73	73	97	5 KP
1E	1C	1C	04	A	4D	74	74	5E	6 Right KP
1F	1B	1B	16	S	E0 4D*	E0 74*	6A	4F	Right CP
20	23	23	07	D	4E	79	7C	57	+ KP
21	2B	2B	09	F	4F	69	69	59	1 End KP
22	34	34	0A	G	E0 4F*	E0 69*	65	4D	End CP
23	33	33	0B	H	50	72	72	5A	2 Down KP
24	3B	3B	0D	J	E0 50*	E0 72*	60	51	Down CP
25	42	42	0E	K	51	7A	7A	5B	3 PgDn KP
26	4B	4B	0F	L	E0 51*	E0 7A*	6D	4E	PgDn CP
27	4C	4C	33	: ;	52	70	70	62	0 Ins KP
28	52	52	34	" '	E0 52*	E0 70*	67	49	Ins CP
29	0E	0E	35	~ `	53	71	71	63	. Del KP
2A	12	12	E1	Shift L	E0 53*	E0 71*	64	4C	Del CP
2B	5D	5C	31	| \	54	84			SysRq
2B	5D	53	53	(INT 2)	56	61	13	64	(INT 1)
2C	1A	1A	1D	Z	57	78	56	44	F11
2D	22	22	1B	X	58	07	5E	45	F12
2E	21	21	06	C	E0 5B	E0 1F	8B	E3	Win L
2F	2A	2A	19	V	E0 5C	E0 27	8C	E7	Win R
30	32	32	05	B	E0 5D	E0 2F	8D	65	WinMenu
31	31	31	11	N	70	13	87	88	katakana
32	3A	3A	10	M	73	51	51	87	(INT 3)
33	41	41	36	< ,	77	62		8C	furigana
34	49	49	37	> .	79	64	86	8A	kanji
35	4A	4A	38	? /	7B	67	85	8B	hiragana
35+	4A+	77	54	/ KP	7D	6A	5D	89	(INT 4)
36	59	59	E5	Shift R	[7E]	6D	7B		(INT 5)

13.6 Transistor

Using a transistor is quite simple, but selecting the right type is a bit more complicated. The method described here was devised by John Hewes[66] of the Kelsey Park Sports College Electronics Club.

There are two basic types of transistors: NPN and PNP. An NPN transistor will conduct when a small positive voltage[67] is applied to the base, which in microcontroller terms is a "1". Only a very small current is required, because the transistor can amplify the current.

In addition to switching and amplifying current, you can also increase the voltage to your device, simply by connecting the transistor to a higher voltage. The ground wires of the microcontroller and the power source of the transistor must in that case be connected to each other. NPN is the type you will be using most, because it seems logical to switch the device on using a "1" from the microcontroller.

The following steps will help select and use the proper NPN transistor:

1. The maximum allowable current through the collector ($I_{c\,max}$) must be larger than the current consumed by the load (I_c).

2. The minimum current amplification (gain) of the transistor $h_{FE}(min)$ must at least be 5 times larger than the collector current I_c divided by the maximum current the microcontroller can deliver.

3. Select a transistor from the table at the end of this section that matches these criteria and make a note of its properties $I_{c,max}$ and $h_{FE}(min)$

4. Calculate the value for the base resistor, R_B, using the formula:

$$R_B = \frac{V_c * h_{FE}}{5 * I_c}$$

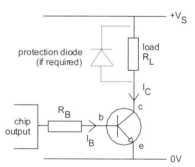

Figure 256 NPN transistor

---

[66] Used in adapted format with permission from the author. For more information see the website http://www.kpsec.freeuk.com/index.htm
[67] Here is an easy to remember rule of thumb applicable to any electronic component: when conducting, the arrow in the symbol always points to the ground lead.

where V_c is the voltage that the microcontroller delivers.

In a simple situation where all voltages are the same you can use the simplified formula:

$$R_B = 0.2 \times R_L \times h_{FE}$$

5. Choose the closest available standard value for R_B.

6. If the load contains a coil (such as a relay or motor) you need to use a protective diode.

The PNP type works just the opposite way, where the transistor will conduct when the base is connected to ground - low in microcontroller terminology. Since this is less intuitive from the microcontroller's point of view - you would expect something to go on when you make a pin high - you will not use a PNP type very often.

Selecting a PNP transistor is done using the same steps as above, except it needs to be connected differently.

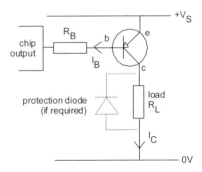

Figure 257 PNP transistor

Table with a few NPN transistors

Code	Type	Case style	I_C max.	V_{CE} max.	h_{FE} min.	P_{tot} max.	Category (typical use)
BC238	NPN	TO92C	100mA	25V	120	500mW	Gen. purp., low power
BC547B	NPN	TO92C	100mA	45V	200	500mW	Audio, low power
BC547C	NPN	TO92C	100mA	45V	420	500mW	Audio, low power
2N3053	NPN	TO39	700mA	40V	50	500mW	Gen. purp. med.power
TIP29A	NPN	TO220	1A	60V	40	30W	Gen. purp. high power
TIP31A	NPN	TO220	3A	60V	10	40W	Gen. purp. high power
2N3055	NPN	TO3	15A	60V	20	117W	Gen. purp., high power

Table with a few PNP transistors

Code	Type	Case style	I_C max.	V_{CE} max.	h_{FE} min.	P_{tot} max.	Category (typical use)
BC177	PNP	TO18	100mA	45V	125	300mW	Audio, low power
BC178	PNP	TO18	200mA	25V	120	600mW	Gen. purp. low power
BC179	PNP	TO18	200mA	20V	180	600mW	Audio (low noise)
BC327	PNP	TO92B	500mA	45V	100	625mW	Gen. purp.
TIP32A	PNP	TO220	3A	60V	25	40W	Gen. purp. high power
TIP32C	PNP	TO220	3A	100V	10	40W	Gen. purp. high power

Explanation of the columns:

Type
This shows the type of transistor, NPN or PNP. The polarities of the two types are different, so if you are looking for a substitute it must be the same type.

Case style
There is a diagram showing the leads for some of the most common case styles. Note that this is a bottom view, with the wires facing you.

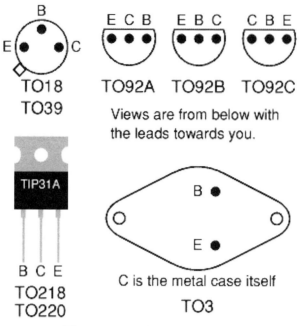

Figure 258. A few pin layout's.

I_C max.
Maximum collector current.

V_{CE} max.
Maximum voltage across the collector-emitter junction.

h_{FE}
This is the current gain (specifically, the DC current gain). The guaranteed minimum value is given because the actual value varies from transistor to transistor - even for those of the same type!

P_{tot} max.
Maximum total power that the transistor can handle. Note that a heat sink will be required to achieve the maximum rating. This rating is important for transistors operating as an amplifier, where the power is roughly $I_C \times V_{CE}$. For transistors operating as switches the maximum collector current (I_C max.) is more important.

Category
This shows the typical use for the transistor.

13.7 Content of the download

In the free download package you will find all software that is used in this book conveniently put together in a single zip file. All software is free, you do not have to pay anything. You don't even have to own this book in order to make use of this package. The software is meant for Windows XP or similar Microsoft operating systems. Follow the installation instructions very carefully! Particularly if you have little or no experience with software installation.

1. Download the free packet using the link op www.boekinfo.tk.
2. Put the file you just downloaded on drive c in the root directory c:\ After you have verified that everything works as it should feel free to move the package to any other location. By starting in the root all software is pre-configured so it will run straight "out of the box".
3. Unzip while maintaining the directory structure - use "extract to here"- you will need about 20 Mb of free space.
4. Run the setup program in the c:\picdev2\tools\vb50setup directory. This will install a number of dll and ocx files that are required for some of the auxiliary programs. You can accept the default directory, or choose something else, it doesn't matter. Before you run this file make sure that its path is indeed c:\picdev2\tools\vb50setup. If it is not you didn't put the package in c:\ but somewhere else. Go back and correct this, otherwise none of the software will work.
5. If you want to connect the Wisp648 programmer to another comport than one, then start JALedit and select the menu options "Compile", "Environment Options", "Programmer". The bottom line in this window is "port com1 wait err go %F". Change the number 1 to the number of the com port you want to use. Note that there is no space between com and the number!
6. Now read the tutorial and discover the exiting world of microcontrollers!

You should have this directory structure on your PC:

```
        <DIR>    JAL
                <DIR>    Compiler
                <DIR>    Libraries
                <DIR>    JALedit
        <DIR>    Project
                <DIR>    lopend
                <DIR>    gereed
    <DIR>       xwisp
    <DIR>       tmp
    <DIR>       tools
```

<DIR>	WinOscillo
<DIR>	Signal generator
<DIR>	Wisp passthrough
<DIR>	Capture frequency
<DIR>	CMUcam2
<DIR>	MICterm
<DIR>	VB50 setup
<DIR>	frequency analyser
<DIR>	Camera take a picture
<DIR>	7segment
<DIR>	WAVconvert
<DIR>	HyperTerminal PE

The main programs within these directories are:

<DIR>   JAL

C:\PICDEV2\JAL\Compiler	This is the directory of the JAL compiler with its support files.
jalv2.exe	This is the JAL compiler. It is called from JALedit, so normally you don't need to do anything with this file. If you want to work from the command line this would be the place to do it.
C:\PICDEV2\JAL\Libraries	This directory contains all libraries discussed in this book, plus some that were discussed in one of my previous books.
xxxxx_bert.jal	The main combination libraries that you need for every project, where xxxxx stands for a particular microcontroller.
adc_hardware, compiler_constants, i2c_sw, random, pwm_hardware, serial_software, pic_program_eeprom, regedit, random et cetera	Some JAL libraries. Credit for these libraries go to their authors.

<DIR>   Project

C:\PICDEV2\Projecten\lopend	This is the "working" directory, which contains projects in progress. It is convenient to use a separate directory for this, which could also contain modified local copies of libraries and Visual Basic source code. In this way the original files remain intact. This directory is empty.
C:\PICDEV2\Projecten\gereed	Completed projects. All of the programs in the book are in this directory, arranged by section number. If you want to modify projects it is perhaps best to copy them to the working directory so that the originals remain intact.

<DIR>   xwisp

C:\PICDEV2\xwisp	This directory contains the program that transfers the compiled source (the hex code) to the programmer.
xwisp.exe	This is the transfer program itself.
xwisp_gui.exe	This is the GUI (graphical user interface) of the transfer program for stand-alone use.
help.txt	An overview of the commands you can use in combination with Xwisp. You only need this if you are not using JALedit.

<DIR>   Tools

C:\PICDEV2\Tools	In this directory you will find a collection of auxiliary programs discussed in this book.
WAVconvert	Program to convert WAV files to other formats such as BTc encoding.
MICterm	Terminal program with functionality specially developed for use with microcontrollers.

VB50 setup	Program to install all dll and ocx files required by the other Visual Basic programs.
Signal generator	A program to generate different kinds of signals with different frequencies.
Resistors	A program for decoding resistor color codes
Capacitors	A similar program for capacitors.
CMUcam2	This directory contaisn the CMUcam2 GUI and the support files. Note that you need Java 1.4 or higher installed on your PC in order to run the GUI.
Frequency analyser	A program to measure the frequency of pulses. It uses your soundcard, so you'll need the special connections as described in section 2.2 Take care to keep the voltage low so that you don't damage your soundcard.
WinOscillo	This program is a simple software oscilloscope. It uses your soundcard, so you'll need the special connections as described in section 2.2. Take care to keep the voltage low so that you don't damage your soundcard.
Wisp passthrough	This small program can switch the Wisp programmer into passthrough mode for use with terminal packages that don't have that capability (for example HyperTerm).
Capture frequency	This program is used in combination with one of the projects to measure the frequency of a signal.
Camera take a picture	This program is used in combination with one of the projects to take pictures on a PC using the CMUcam2.
7segment	This program can make numbers and letters for a 7-segment display.

| HyperTerminal PE | A slightly more advanced version of HyperTerm. You need this program for use with Windows Vista and 7 because they come without HyperTerm. |

<DIR>   JALedit

C:\PICDEV2\JAL\JALedit	In this directory you will find JALedit and its supporting files.
JALedit.exe	This is JALedit. It is perhaps a good idea to make a link to it on your desktop, because this is the program you will be using most. Of course you can also use another editor, or work from the commandline.

13.8 Tips and tricks

This is a small selection of handy tips and tricks.

1. Wait for a pin change

Suppose you want to wait until a switch is no longer engaged.

    while switch loop end loop

2. Combine bits to make a byte

When you want to combine a number of bits to a byte you can do that by declaring them at a certain location of that byte. Lets assume you have a byte called *Demo* and you want a bit variable called *dPC0* to be at the first bit of this byte, you can use this statement:

    var volatile bit dPC0 at Demo : 0

In this way you could declare the entire byte in bits:

    -- declare the variable
    var byte Demo

    -- declare which bit goes where
    var volatile bit dPC0 at Demo : 0

```
var volatile bit dPC1 at Demo : 1
var volatile bit dPC2 at Demo : 2
var volatile bit dPC3 at Demo: 3
var volatile bit dPC4 at Demo : 4
var volatile bit dPC5 at Demo : 5
var volatile bit dPC6 at Demo : 6
var volatile bit dPC7 at Demo: 7
```

Now if you set one of the bit variables to 1, like this:

```
dPC5 = 1
```

Then *Demo* has gotten the value 0b_0010_0000 or 32 (assuming the other bits were still 0). Of course this also works the other way around; if you give *Demo* the value 4 *dPC2* will be 1.

3. Take a byte apart to bits

You can remove bits from a byte one by one. Declare a bit variable, such as *codec*, as the leftmost bit of a byte, such as *value*. Then shift the byte left by one position 8 times to make *codec* get all of the bits of *value* one by one, starting at the high bit.

```
var byte value
var bit codec at value : 7

for 8 loop
  led = codec
  value = value << 1
end loop
```

4. Input or output pin on the fly

In the beginning of your program you need to declare whether a pin will be an input or an output. But that doesn't mean it has to stay that way. While the program is running (on the fly, so to speak), you can easily change the direction of a pin. For example, when you are communicating over a single wire:

```
if pin_a5 then
    -- signal from the other pic
    -- wait until its low again
    while pin_a5 loop end loop
    -- switch to output and make high
```

```
         pin_a5_direction = output
         pin_a5 = true
         -- wait 10 seconds and make low again
         delay_1s(10)
         pin_a5 = false
         -- make the pin input again
         -- and wait for the next instruction
         pin_a5_direction = input
     end if
```

5. Pin that doesn't go high

Pin RA4 is a so-called open collector pin. This means it cannot be made high. The best thing to do is use a pull-up resistor (a 10k resistor to +5 V). That way the pin is always high, unless you make it low. If you need more power you can reduce the resistor value a bit.

The 12F675 doesn't have this problem, because it doesn't have an RA4 pin. It does have an MCLR pin, which can only be used as an input.

6. Combined operators

The operators that were discussed in section 13.1 can also be used combined, for example in an *if..then..* command. For example:

    if ((a > 4) & (b < 3)) then

In this case the *then* clause will be executed if a is greater than 4 and at the same time b is smaller than 3

7. Calculate with bytes and words

You would perhaps expect that this program:

```
    var byte a,b
    var word c

    c = a * b
```

would give the right answer for all values of a and b, for the product of two bytes always fits in one word. In practice that isn't the case. The compiler will calculate the right side of the equation first, and as an intermediate variable it will use the largest variable type on that side. In this equation the largest type on the right side is a byte, so bytes a and b will be multiplied in a byte, and when that is done the result is transferred into word c. But by then it is already to late, and anything larger that a byte is lost. In order to prevent this you can use the word classifier:

    var byte a,b
    var word c

    c = word(a) * word(b)

Index

!	70, 87, 393
!!	393
!=	394
%	106, 393
&	132, 394
^	394
_usec_delay	73
|	394
<<	393
==	394
>>	200
>>	133
0b	89
1_wire	410
10f200	371
16f648A	362
16f877A	14, 29, 110
16f877A option_reg	83
18f4455	14, 95
18f4455 intcon	104
18f4455 t0con	102
18f4685	14, 114
1N4007	46
1-wire	172
240 volt	59, 78
39MF22l	56
7segment	342, 414, 426
A/D	74
A/D channels	400
A/D converter	400
A/D high resolution	169
ADC	74
ADC_hardware_Nchan	400
ADC_init	400
adc_low_res	74
ADC_off	75, 400
ADC_on	75, 400
ADC_read	169
ADC_read	75
ADC_read	400
ADC_read_bytes	75, 400
adc_read_low_res	166
ADC_read_low_res	75, 400
aluminium foil	158
amplifier	109
AN	74
Analog input	31
AND	132
array	100, 392
Array	392
ascii codes	415
assembler	44, 387
at	200
attention mode	185
batch file	185
BC547	347
BC547C	208
BF	259
binary	88
bistable	50
bit	54, 378
bit at	380
bit mask	314
bitlength	378
bitrate	115
block	343
Block	391
bouncing	352
breadboard	11
brgcon 18f4685	239
buzzer	164
byte	322
byte	54
byte	378
byte calculation	329
byte in	73, 292

byte is	380
byte out	293
byte out	384
c1out	125, 137
c3	95
c4	95
c5	95
Camera take a picture	426
CAN buffers	241
CAN bus	237
cancon	238, 250
canstat	240
capacitor	158
capacitors	26
Capacitors	426
capture	140
Capture frequency	426
Capture Frequency	147
carrier	194
Case	343
Case	391
CCP1	96
ccp1con	140
ccp1con 18f4455	206
ccp1if	140, 144
CharData[]	409
ciocon	241
class	214
clock crystal	277
cmcon	124
cmucam2	413
CMUcam2	327, 333, 338, 426
CMUcam2 camera	303
code area	120
coil	46
COM port, virtual	221
comment	395
common anode	342
comparator	123, 135, 143
compile	39
confidence	331
const	191
const	100
const	381
const pwm_frequency	399
const Serial_SW_Baudrate	337
constant	54
Constants	381
Cool Edit	121
crystal	31, 83
CUI	308
cut-off frequency	94
CVR generator	130
cvrcon	130
D/A	298
D/A	92
d1w_present	180, 410
d1w_read_bit	175, 180, 410
d1w_read_byte	180, 410
d1w_read_byte_with_CRC	179, 411
d1w_reset	180, 410
d1w_write_bit	175, 180, 410
d1w_write_byte	180, 410
data_buffer	224
Data_Eeprom_Read	401
Data_Eeprom_Write	401
Data=Data_Eeprom	401
datasheet	29
debugging	41
decimal	88
decoupling capacitor	12
delay_1s	49
delay_1s(1)	38
delay_37uS	73
device	214
dice	227, 229
dice	403
dielectric	158
digital in-/output	31
DIP05-1A72-12L	45
direction	49
DoEvents	188
dong ding	223, 228
ds1307	413

DS1307	271, 281, 285
ds1307_entertime	277, 413
ds1307_readtime	277, 413
ds1307_sendtime	277, 413
ds1307_showtime	277, 414
DS1822	176
DS1822_read_temperature_raw	179, 411
DS1822_start_temperature_conversion	179, 411
ds1882_1_wire	410
duty cycle	66, 93
dword	54, 378
ear, artificial	135
eeprom_ackpoll	407
eeprom_write	407
Electret	131, 135, 148
elsif	390
enable_modem	405
Enable_Modem	216
end task	389
eprom_read	407
Esc commando's	415
Esc commands	190
Excel	235
Exit loop	381
falling edge	140
FD	331
finger	155, 159
firmware	361
flag	87
for loop	73
For loop	382
Forever loop	381
Frame diff	309
frequency	142
Frequency analyser	426
frequency generator	28
Function	385
gain	93
gate	76
gate	62
gie	403

GM	313, 326
Grab Frame	308
gv	306
Hall effect	150
hardware serial	396
hexcode	44
h_{FE}	419
HID	223, 230
hid_keyboard_byte	235, 406
hid_keyboard_key	235, 406
hid_mouse_write	225
HID_Mouse_Write	229, 407
host	214
HTPE	189
hub	214
human sensor	154
HyperTerm	181, 189
HyperTerminal PE	427
i2c	414
I^2C	270
i2c_get_data(	273, 414
i2c_put_ack	273, 414
i2c_put_data	273, 414
i2c_put_nack	273, 414
i2c_put_read_address	273, 414
i2c_put_stop	273, 414
i2c_put_write_address	273, 414
i2c_sw	407
i2c_wait_ack	273, 414
i2cp	414
ID number	173
identifier bits	242
If then else	389
if..then..	54
I_{hold}	63
include	118, 377
infrared	194
infrared LED	207
initialising	54
In-line assembler	386
intcon	358, 403
interrupt	356

interrupt	83, 103
interrupt	385
interrupt frequency	84
interrupt frequentie table	83
interrupt procedure	85
intf	358
iocon	288
iodir	288
JAL	15
JALedit	18, 29, 35, 427
Java	307
keyboard scan codes	231
keyboard send report	230
keycodes	231
laser	364
lcd_44780	408
lcd_44780	408
LCD_blink	409
LCD_char_line_pos	408
LCD_char_pos	408
LCD_clear_line	408
LCD_clock_line_pos	409
LCD_cursor	409
LCD_cursor_pos	409
LCD_custom	409
LCD_display	409
LCD_high_low_line_pos	408
LCD_init	408
LCD_num_line_pos	408
LCD_num_pos	408
LCD_num_pos_1dec	408
LCD_progress	408
LCD_shift_left	409
LCD_shift_right	409
LDR	364
LED	47
Left shift	393
LF	331
library, make your own	293
light dimmer	71, 78
lithium button cell	277
LM317	265, 303
LM317	164
LM336	163
LM35	168
LM358	148
Load_My_ID	179, 411
long array	100
Long table	392
lookup table	100
Lookup table	392
loudspeaker	90
low pass filter	93
LUT	100, 118, 392
Lux	364
main terminal	62
make codes	231
Marchaudon	171
mask	244
master	172
MatchRom	179, 411
math	170
MAX517	298
max517_write	298
MCLR#/THV	47
MCP23008	287
MCP2551	246
MEB-6	151, 164
microphone	130, 135
MICterm	20, 127
mips	15
mm_mchpusb.sys	221
mmc	268, 412
MMC card	263
mmc_init	268, 412
MMC_read_close	268, 412
MMC_read_open	268, 412
MMC_read_streaming_close	268, 412
MMC_read_streaming_open	268, 412
MMC_write_close	268, 412
MMC_write_open	268, 412
MOC3020	63
MOD	393
modulo	106

mosfet driver	51
mouse	223
mouse report	224
MPLAB	44
MSComm	187
multimeter	300
normalized	100
not	70
NPN	419
Ohm's law	45
OM	313
opamp	148
Operators	393
option_reg	357
optocoupler	63
OSC1/CLKIN	47
OSC2/CLKOUT	47
P=V*I	52
passthrough	127
PDIP	15
peak voltage AC	64
PICterm	425
PID	215
pin a4	67, 125, 143
PM	312, 326
PNP	419
police	112
port B interrupt	356
potmeter	74
power supply	303
pr2	204
Pragma	394
pragma clear	395
pragma eedata	395
Pragma EEDATA	401
pragma interrupt	85, 395
pragma task	389, 395
pragme interrupt	385
prescaler	83
procedure	292
procedure	73
Procedure	383
Program_EEprom	401
Program_EEprom_Read	401
Program_EEprom_Write	401
protection diode	46, 164
pull-up	177, 195
pulse width modulation	66
PWM	66, 93, 114, 140
PWM duty cycle	207
PWM period	203
PWM_init_frequency	98, 399
pwm_set_dutycycle	106
PWM_Set_DutyCycle	98, 399
pwm1_dutycycle	399
pwm2_dutycycle	399
RAL 5W-K	51
random_BYTE	403
random_WORD	403
RC filter	104
RC network	93
RC network debouncing	352
read_ID	179, 411
reedrelay	45
reference voltage	129
regedit	410
register	83
register_debug	410
register_read	410
register_write	410
remote control	210
Repeat until	383
reset on pin 1	281
resistors	26
Resistors	426
Right shift	394
rising edge	140
rotary encoder	351
RS232	183
RW	315
rxm0sidl	244
sampling rate	115
sbyte	54, 378
Schmitt trigger	123

SDI	253
sdword	54, 378
segnumber	346, 414
Send_ID	179, 411
serial port	183
Serial_HW_Baudrate	398
Serial_HW_Byte	398
Serial_HW_Clear	398
Serial_HW_Data	398
Serial_HW_Home	398
Serial_HW_Locate	398
Serial_HW_Printf	398
Serial_HW_Read	320
Serial_HW_Read	320
Serial_HW_Read	398
serial_hw_write	321
Serial_HW_Write	320, 398
Serial_SW_Baudrate	397
Serial_SW_Byte	128, 184, 397
Serial_SW_Clear	128, 184, 397
Serial_SW_Home	128, 184, 397
Serial_SW_invert	397
Serial_SW_Locate	128, 184, 397
Serial_SW_Printf	128, 184, 397
Serial_SW_read	128, 397
Serial_SW_Read	184, 187
Serial_SW_read_init	397
Serial_SW_write	128, 397
Serial_SW_Write	184
Serial_SW_write_init	397
Serial_USB_Byte	216, 405
Serial_USB_Clear	216, 405
serial_usb_disable	216, 217, 406
Serial_USB_Home	216, 405
Serial_USB_Locate	216, 405
Serial_USB_Poll	216, 405
Serial_USB_Printf	216, 406
Serial_USB_Read	216, 405
Serial_USB_Write	216, 405
SF	336
Sharp	194
signal generator	28
Signal generator	426
Signal Generator	147
sinus	100
slave	172
slave select	253
snubber	59
software oscilloscope	23
solid state	56
sound threshold	129
Spectrum Analyser	92
speech	114
SPI	253
SPI mode	254
spi_hardware	411
spi_software	411
spi_transceive	259, 260, 267, 411, 412
SS	253
sspbuf	259
sspcon1	258
sspstat	256
start task	389
static electricity	155
suspend	389
sword	54, 378
t0ie	403
t0if	86, 106
t1con	140
t2con 18f4455	205
Task	388
TC4427A	51
TDA1011	109
television	199
thermometer	171
third hand	304
threshold	123
TIC206M	63
time constant	352
time registers DS1307	272
timer0	83
timer0 interrupt 16f877A	83
timer0 interrupt 18f4455	102
timer0 interrupt 18f4685	115

timer1 140
tmr1l 144
TMR2prescaler 204
touch key 154
transformer 58
transistor 208, 347
transistor 419
triac 62, 67
true 69
TSOP1737 .... 195, 203
TSUS 5202 207
twisted pair 247
two's complement 177
txbndlc 242
UA7805 12
UGN3140u 150
USART_HW_Serial 398
USB 214
USB driver 221
USB plug 218
usb_delay_100ms 229, 407
usb_delay_1ms 229, 407
usb_hid_keyboard 406
usb_hid_mouse 407
usb_initialized 229, 407
usb_is_configured 229, 235, 406, 407
usb_rs232 405
USB_Tasks 229, 235, 406, 407
V=I*R 45
variable 54
VB50 setup 426
V_{DD} 47
V_{gate} 63
VID 215

Visual Basic 185
Visual Basic 22
volatile 380
voltage regulator 164
voltagereference 163
V_{SS} 47
VT52 189
VW 312, 326
wait_for_colon 321, 413
wait_for_packet 320, 321, 413
wait_for_reply 321, 413
watch dog timer 83
wav file 114
WAVconvert 122, 425
WDT 83
while loop 69
While loop 382
Windows device manager 222
WinOscillo 426
WinOscillo 68, 99, 104
WinOscillo bus 247
Wisp connection 370
Wisp pass through 426
Wisp628 13
Wisp648 361
Wisp648 50
Wisppassthrough 181
WispPassThrough 185, 193, 279
word 54, 378
word(byte) 349
xwisp 425
xwisp_gui 363
zero crossing 67, 72

437